AF386182

Heinz Guderian and the Rise and Fall of Hitler's Tank Units

Heinz Guderian and the Rise and Fall of Hitler's Tank Units

Jaap Jan Brouwer

Pen & Sword
MILITARY

First published in Great Britain in 2026 by
Pen & Sword Military
An imprint of Pen & Sword Books Limited
Yorkshire – Philadelphia

ISBN 978 1 03614 101 1

A CIP catalogue record for this book is available from the British Library.

Typeset by Mac Style
Printed in the UK by CPI Group (UK) Ltd, Croydon, CR0 4YY.

The Publisher's authorised representative in the EU for product safety is Authorised Rep Compliance Ltd., Ground Floor, 71 Lower Baggot Street, Dublin D02 P593, Ireland.
www.arccompliance.com

For a complete list of Pen & Sword titles please contact:

PEN & SWORD BOOKS LIMITED
47 Church Street, Barnsley, South Yorkshire, S70 2AS, England
E-mail: enquiries@pen-and-sword.co.uk
Website: www.pen-and-sword.co.uk
or
PEN AND SWORD BOOKS
1950 Lawrence Road, Havertown, PA 19083, USA
E-mail: uspen-and-sword@casematepublishers.com
Website: www.penandswordbooks.com

Contents

Chapter 1

The Early Years

Heinz Guderian was born on 17 June 1888 in Kulm, West Prussia, the son of a German infantry officer. He came from the traditional background of German officers: his grandparents on both sides were landowners in the east of Germany, as it then was. His upbringing in this military family had a great influence on his character. He spent most of his youth in garrison towns, and it is no wonder that he wanted to follow in the footsteps of his ancestors. When he was twelve he was sent with his ten-year-old brother to the cadet school in Karlsruhe, in Royal Baden. His father was at that time stationed at St Avoid, in occupied Alsace-Lorraine. Guderian spent two years in Karlsruhe and in April 1903 moved to the Hauptkadettenanstalt in Gross-Lichterfelde, near Berlin. There he was educated in the Prussian military tradition, which emphasized flexibility, resourcefulness and initiative. He learned quickly and was always ranked among the best in his class.

He himself described his education as a combination of 'military rigour and simplicity … but based on kindness and justice'. Even when measured against the educational climate of the time, the regime at German cadet schools was decidedly spartan. Officer training began in Germany much earlier than, for example, in Great Britain and the United States, where candidates could only enter at the age of eighteen. The teachers were officers, and the pastoral staff were non-commissioned officers of the Imperial Army (*Kaiserliches Heer*). Guderian thrived in this environment, however, since he was strong and exceptionally intelligent. He showed early on the perseverance, independence and ambition that were so characteristic of him in later life. His stubborn self-will came to the fore when, during his final examination, he rejected the prescribed solution to a problem and presented his own answer – which, however, failed to impress the examiners. This forceful demeanour was what would make him one of the most striking personalities in German military history.

After successfully completing all stages of cadet training, he served as an ensign in the 10th Hannover Jägerbataillon in Goslar, then commanded by his father; on 27 January 1908 he was promoted to second lieutenant. He enjoyed the pleasures that came with his officer status: hunting, riding, theatre visits and dances, all activities that followed the family tradition. Guderian fell in love with Margarete 'Gretel' Dörne, but her father, a garrison doctor, considered him too young to marry; this prompted Guderian to wish to leave the oppressive Goslar as soon as possible. He considered taking additional training in machine gunnery, but his father, always his great role model, saw more future in the field

Guderian as a little recruit in the uniform of his father's unit, the Hannoverschen Jäger-Bataillon 10.

of modern means of communication, a recent innovation. The young man's choice ultimately fell on this subject, which would have a great influence on his later military career.

His experience with the Jägerbataillon also played a major role in his future. These Jägerbataillons – light infantry – fulfilled the role of reconnaissance units (equipped with bicycles) for an infantry corps in the Imperial Army, or of support troops for a cavalry division. The Jägerbataillons could be seen as mobile units, comparable to the rifle regiments of the British Army or the *chasseurs* of the French Army. They tended to regard the 'heavy' infantry as slow and cumbersome. Another important characteristic of these reconnaissance units was that their men, who were recruited as much as possible from forested and mountain areas, were taught to act independently in battle, something expected much less of soldiers in ordinary line regiments.

After completing his communications training, Guderian joined the 3rd Telegraf Battalion in Koblenz on 1 October 1912. Here he also took part in preparatory training for the Kriegsakademie and learned to speak fluent French and English. In 1913, at the age of twenty-five and the youngest in his year, he managed to pass a competitive examination at the Kriegsakademie. An education at this higher military school was essential for an officer who wanted to achieve success in his career. There was fierce competition for it, and it was a considerable achievement for Guderian to have been selected at such a young age. This was not surprising: Guderian had a great interest in military history and an excellent memory, which would serve him well during his later career. He focused so much on his studies and professional development that other activities often fell by the wayside, and he had few friends. One of the

Guderian, on the left, in front of the Jäger Memorial during his training at the Kriegsschule in Metz, 1907.

few was another candidate officer, Bodewin Keitel; the Keitel family would play a significant role in his life. His intelligence and speed of thought earned him the nickname 'Schnelle Heinz' from his fellow students. One of his mentors whom Guderian greatly admired was Colonel Count Rüdiger von der Goltz, an energetic and charismatic senior officer; his and Guderian's paths would cross again after the First World War.

His admission to the Kriegsakademie had meanwhile dispelled the objections of his intended parents-in-law: on 1 October 1913, Heinz Guderian and Gretel Dörne were married. Gretel had a down-to-earth view of life and would be a moderating influence on her husband in the years to come.

The training at the Kriegsakademie was disrupted by the outbreak of the First World War, whereupon suitable staff positions were sought for all students. Since Guderian had worked for a time in a communications battalion before he came to the Kriegsakademie, it was obvious that he would be given command of a radio station. When the 10th Jäger Bataillon marched west in August 1914 as part of the Third Army, Guderian took command of the radio station at the headquarters of the 5th Kavallerie Division. This responsible position showed that he had already earned his spurs as an officer. The experience he gained during the rapid advance through the Ardennes made a great impression on him and, twenty-six years later, resulted in his conviction that this manoeuvre could be repeated. During the advance he demonstrated his characteristic style of leadership in mobile operations: he was always at the front, often came under fire and narrowly escaped capture. When the front came to a standstill, Guderian recognized the advantages of aircraft for gathering information; he was one of the few officers who gained experience of aerial reconnaissance in 1914. The first – unsuccessful – attempt on 22 April 1915 to force a breakthrough on the Ypres front by using mustard gas would then ensure that in his later career he would always emphasize the danger of premature deployment of new weapons.

When his service with the 5th Kavallerie Division ended later in April 1915, Guderian became assistant liaison officer at the headquarters of the Fourth Army. This in itself was a remarkable position. It was customary

Guderian as Oberleutnant in December 1915 on the staff of the Fourth Army.

in the German Army to have officers with staff training regularly switch between a staff job and a command in the field. It was therefore expected that Guderian would return to his regiment, for example as a company commander. However, he continued to hold the post of assistant liaison officer until April 1917 and, promoted in the meantime, was then posted to the staff of the 4th Infanterie Division. From that moment until the end of the war he would only hold staff positions. Finally, in January 1918, he was given the opportunity to complete his training, interrupted by the war, in Sedan, and on 28 February he was appointed to the Generalstab (General Staff) of the *Kaiserliches Heer*.

Two innovations that would be important in the decades to come had meanwhile been introduced, largely unnoticed, at the front. Guderian was in Flanders when the British first deployed tanks on 15 September 1916. Like many officers, he was not impressed; he regarded the tank as one of many new weapons and tactical adaptations that both sides were using in an attempt to break the deadlock at the front. One of the German adaptations was the

French Chaumond tanks. This type of tank, in fact the first mechanized and armoured gun, had problems crossing trenches and ditches due to its Holt-type tracks, as can be seem in the picture.

French Schneider tanks are driven off a train.

'delaying defence', a well-thought-out and extensive system in depth that proved to be an extremely efficient solution when on the defensive. Guderian, with his strongly offensive attitude, would in later years reject this doctrine as reactive, only to advocate it himself as a field commander in the Soviet Union in the last months of 1941, when the German offensive on Moscow had stalled. After initially unimpressive attempts, the Allies succeeded in breaking through the Hindenburg Line with tanks at Cambrai in a single morning on 20 November 1917. This underlined the strategic potential of the tank on the battlefields of the future. There is no evidence, however, that Guderian showed more than a superficial interest in them at the time.

His attention was focused more on another innovation on the German side: the *Hutier Taktik*. In order to deal with the problem of breaking through the front, the Germans developed this tactic in the final phase of the First World War: heavily armed units (*Stosstruppe*, shock troops) infiltrated through the front line at certain key points (*Schwerpunkte*) and then rapidly advanced into the enemy's rear to disrupt communications and destroy artillery positions. The tactic was named after the commander-in-chief of the German Eighth Army in Russia in 1917, General Oskar von Hutier. The high command of

A Canadian armoured personnel carrier during the Battle of Amiens.

the German Army ordered him to apply the new tactic developed by the General Staff to the fortifications of Riga, which had resisted a two-year siege by the Germans. Using this tactic, Riga was taken within two days, and from that moment it was known as '*Hutier Taktik*'.

On 21 March 1918, during the Easter offensive, the Germans first used *Hutier Taktik* on a large scale in a final attempt to defeat the British-French armies before the American Expeditionary Force reached the front. It proved to be a great success, and the Germans rapidly regained the ground they had previously lost. The combination of *Auftragstaktik* (see Chapter 5), tanks and a well-thought-out communications network would form the basis of the success of the Blitzkrieg twenty years later, as we shall see. But all this was ultimately to no avail, and the German Army was forced to halt its offensive in August 1918 and withdraw to positions behind the front. Guderian's mission during this period was to supply the front units, but despite all their efforts, they had to retreat to the Siegfried Line by mid-September. Shortly afterwards, Guderian was transferred to the German mission attached to the Austro-Hungarian Army fighting in Italy. Soon after he arrived in Italy, the Austrians suffered a crushing defeat at Vittorio Veneto which forced Austria to pull out of the war as quickly as possible.

With nothing left to do in Italy, Guderian returned to Berlin in November, where the spirit of revolution had been let out of the bottle and the Kaiser was forced to flee into exile. As the front collapsed, the provisional government called for an armistice, which was finally concluded on 11 November. In the weeks that followed, the German Army retreated in remarkably disciplined fashion, and it was assumed that after the disorder that followed the end of hostilities, the loyalty of the army would be restored. Upon arrival in Germany, though, returning soldiers simply left their barracks, either alone or in small groups. They headed back to their homes and families, ignoring orders to stay put. By the end of December the exodus was complete and the new Weimar Republic effectively no longer had an army. In the meantime, social revolution was in the air. In Berlin, civil war was looming, and Munich, the capital of conservative Bavaria, was in the hands of a socialist government, which was immediately branded as a 'bolshevist' by the Prussian middle class. Unrest from left and right was threatening the new state, which urgently needed a loyal and effective army to prevent lawlessness from taking hold and further destabilizing Germany. Guderian was shocked to see how quickly social and political disorder had spread and how quickly the military had lost its prestige. In particular, the rapid spread of Marxist ideas among the soldiers would change his worldview forever. To make matters worse, the Polish and Soviet armies were advancing towards the eastern borders of Germany, the lands where his family had lived for centuries.

During this period, the General Staff found itself in the strange situation of having no army to exercise authority over. Although there were still half a million soldiers under arms, they were stationed in the areas that had been taken from the Russians under the Treaty of Brest-Litovsk, between the areas conquered in Ukraine, where the majority of these units were stationed, and the German border lay a large belt where no authority prevailed. Part of it had belonged to Austria, which had collapsed two months earlier, and part to Germany, which was, however, unable to assert its authority there. Other parts were Russian, but the Russians, and later the Soviets, had been driven out of these. The rest of Eastern Europe was up for grabs, and many nationalities were lurking to strike: Poles, Czechs, Russians, Germans, Estonians, Lithuanians, Latvians, French and British. The Balkans were destabilized, a civil war was raging in Russia and Ukraine had its own separatist movement.

While the borders of a future Europe were being drawn at the negotiating table, with Poland as an important buffer state, the so-called Freikorps emerged from the *Kaiserliches Heer* in Germany. These were units of hardened front-line soldiers who had managed to maintain their cohesion in one way or another. Some had nationalist or counter-revolutionary motives, others rallied behind a charismatic leader. These Freikorps were also active in the east. The soldiers in these units ignored the government's order to withdraw from the area assigned to the Soviets and the Poles. Instead, they formed local alliances with anti-Bolsheviks, Baltic nationalists and White Russians. The General Staff and the High Command of the Army were ambivalent towards these Freikorps, which filled the vacuum in the east as long as no newly formed force was available. In addition, in the eyes of both the German government and the Allies, the anti-communist Freikorps formed a welcome shield against the advance of communism. In this atmosphere, Guderian was linked in March 1919 to the Zentralstelle Grenzschutz Ost, around which most Freikorps had gathered to protect Prussia against the advancing Soviets.

One of the central units within the Freikorps was the Eisernen Brigade under the command of General Rüdiger von der Goltz, Guderian's old mentor. This brigade covered the retreat of the fleeing Germans to Riga and formed an important instrument in German strategy towards the Baltic states. Having grown into the Eisernen Division, it launched an offensive towards Lithuania and Latvia, while other Freikorps in Silesia and Prussia tried to prevent Polish units from occupying historically German territory. Guderian was attached to the staff of the Eisernen Division in June 1919, with the idea that in this way the General Staff could strengthen its hold on the formation. The Eisernen Division was at that time about to occupy Riga, the main port of Latvia, after having driven the Soviets out of the region.

However, the representatives of Latvia found a sympathetic ear among the Allies to their warnings that the ultimate goal of the Germans was to establish a puppet government in a state that, according to President Wilson's principles, should be allowed to decide its own future on the basis of 'national self-determination'. The Allies, who had until then, in accordance with the terms of the armistice, urged the Germans to keep the peace in the east (i.e. to prevent the Bolsheviks from conquering too much territory), began to become concerned about the Germans' intentions. Diplomatic and maritime activities forced the Germans to suspend their operations, and the Latvian Army was reinforced. In addition, the new Weimar government had signed the Treaty of Versailles on 28 June, and Hans von Seeckt, the new commander of the army, wanted the German units to comply with the Treaty. The Eisernen Division, however, did not accept this and declared that it was no longer part of the Reichswehr. This put Guderian in a dilemma: he had to choose either loyalty to the army and the German state or remain attached to a division that, like him, rejected the 'Diktat' of Versailles. Guderian leaned towards the latter course and stood firmly behind von der Goltz when he ignored explicit orders to return to Germany on 23 August.

His superiors were 'not amused' by Guderian's attitude, although many of his fellow officers on the General Staff shared his views. He was given

Units of the Eisernen Division on the Baltic Sea on the eve of the failed battle against the Latvian Army. A swastika can be seen on the truck.

a second chance to prove where his loyalties lay when, a few days later, he was formally recalled by the High Command. Guderian obeyed, returned to Germany and in September 1919 was transferred to Brigade 10 of the newly formed Reichswehr in Hanover, far away from the military actions in the east. The Weimar government wanted to maintain a 'temporary Reichswehr' as long as the Allies did not clearly indicate their ultimate wishes. This 'temporary Reichswehr' was built up along the lines of the old *Kaiserliches Heer* and consisted of the old twenty-four Wehrkreisen (military districts). The strength of the units was reduced from corps to brigade level. In January 1920 Guderian was transferred to the 10th Jägerbataillon in Goslar, his father's old unit. He would discover in the coming years, however, that the General Staff had not forgotten his hesitancy in 1919.

By then, all German units had withdrawn from the Baltic Sea area, and Germany had had to abandon its attempts to extend its influence further to the east. It would have to reconcile itself to an Eastern Europe with three independent Baltic states, as well as a Polish state that included a large area that had previously been German. In addition, East Prussia was separated from the rest of Germany in order to give the Poles a corridor to the sea at Danzig.

But the country would remain restless for a long time: in March 1920, the remnants of the Freikorps under von der Goltz briefly seized power in the so-called Kapp Putsch. But now Guderian and other officers, however much they detested this government which they believed had sold Germany out, openly distanced themselves from these and other half-hearted attempts to destabilize the Weimar Republic. His unit was deployed to suppress unrest in the area around Hildesheim and in the Ruhr. In the meantime, Guderian had plenty of time to let his experiences during the First World War sink in. Because of the staff jobs he had held, he had not directly experienced the massacre in the trenches of the Western Front. He was therefore able to scrutinize the course of events with some detachment and to analyse the character of the war and the reasons for the Germans' defeat. He also realized that static warfare in any future conflict had to be avoided at all costs, and in the next years he would slowly come to understand that tanks and armoured vehicles would play an important role in achieving this.

Chapter 2

The Development of Tank Doctrine in the 1920s

While the new Reichswehr was trying to meet the demands of the Treaty of Versailles, Guderian, dedicated as ever, ensured in no time that his Jägerbataillon was in tip-top condition. But the relative peace was about to be disturbed. In January 1922, Lieutenant Colonel von Stülpnagel of the Truppenamt (troop office) of the Reichswehrministerium called Guderian to ask why he had not yet reported to Munich. It turned out that he had been transferred as a staff officer to the Inspektion der Kraftfahrtruppen (motorized troops) of the Reichswehrministerium. The head of the Kraftfahrtruppen, General von Tschischwitz, had requested an officer to supplement his staff. Guderian was, to say the least, surprised by this appointment, which was certainly not in his expected career path. In his memoirs he explained it as a result of being one of the few staff officers with experience of communications and technical training. In reality, it was evidence of the long arm of the General Staff, which had not yet forgotten that Guderian had struggled with his loyalty in 1919. However, his experience did mean he was eminently suited to the position. At that time, no one could have imagined that this 'backroom' appointment would prove to be a blessing in disguise, and that, as a result, Guderian would fill a central role in the build-up of German armoured forces during the coming decade.

By mutual agreement, it was decided that Guderian would first gain experience at the 7th Kraftfahrabteilung in Munich, one of seven such Reichswehr units responsible for supplying their respective military districts; his appointment would then become effective on 1 April. At the 7th Kraftfahrtabteilung he met for the first time Major Oswald Lutz, the commander of this battalion, who would play a decisive role in Guderian's future career. He found a kindred spirit in Lutz, and their collaboration would bear fruit in the coming years. In the 1920s, Lutz would play a prominent role in the field of armoured vehicle development. Guderian gained considerable experience in Lutz's unit and rewrote in passing a number of what he considered to be dry regulations with titles such as *Truppentransport auf Lastkraftwagen* (troop transport by lorry), an exercise that would result in his book *Achtung-Panzer!* and his later publications such as *Tigerfibel* and *Pantherfibel*.

Once on the staff of the Kraftfahrtruppen at the Reichswehrministerium in Berlin, Guderian was very surprised by the work he was assigned, which related to maintenance workshops, fuel storage, construction and communication matters. He protested in vain to von Tschischwitz, on the grounds that this was not the work he had been promised and that he lacked the knowledge and experience to carry it out. His protests fell on deaf ears, and in desperation he asked for a transfer to his old Jägerbataillon in Goslar. But this, too, was rejected

After this setback, Guderian reluctantly set to work on unfamiliar tasks. He learned quickly, however, and one of the most interesting things he

Guderian (right) with Reichswehrminister Gessler during the Hindenburg manoeuvres in Mecklenburg, 1925.

encountered was a study by von Tschischwitz on the possibility of transporting troops by motorized vehicle. The study was based on an exercise in the Harz Mountains and it brought Guderian into contact for the first time with the possibility of deploying motorized units. The First World War had seen examples of motorized transport being used in place of horse-drawn vehicles. However, such troop movements had always taken place behind a relatively static front line; they had never been used in a war of movement. Now that, due the Treaty of Versailles restricting the size of the German military, all of Germany's borders could not be defended simultaneously, it was unlikely that any possible future conflict would be fought from behind a static defence line; there simply were not enough soldiers for that. Instead, Germany would have to adopt a mobile method of defence. In addition to the problem of transporting troops in a mobile war, there was the difficulty of protecting such transports. This could only be done by armoured vehicles. In order to find out more about such vehicles, Guderian contacted Lieutenant Ernest Volckheim, who at that time was attached to the General Staff and was researching the use of German and Allied tanks and armoured vehicles in the First World War.

German theoreticians

Ernst Volckheim was not just anyone: in the 1920s and 1930s he was one of the most important theoreticians in the use and development of armoured vehicles. He was also one of the few German officers who had gained front-line experience with tanks in the First World War. He was posted to the German tank corps in February 1918, and from April to 10 October, when he was wounded in the head, he experienced several tank battles as an officer of the 1st Schwere Panzer Kompagnie, equipped with the German A7V tanks. After the war he was attached to the Inspektion der Kraftfahrtruppen of the Reichswehrministerium in Berlin. There he was given an explicit assignment to further investigate the possible uses of armoured vehicles. In the 1920s, Volckheim was also one of the leading authors in the field of tanks and tank tactics. In 1923, his book *Die deutschen Kampfwagen im Weltkriege* was published, a well-documented history of the German experience with tanks in the First World War. He also wrote more than a dozen articles on tanks and tank tactics for the *Militär Wochenblatt* magazine and the editorial for all issues of *Der Kampfwagen*, a magazine that appeared as a special supplement to the *Militär Wochenblatt* during 1924 and 1925 and was entirely devoted to tanks, armoured vehicles and motorization. His articles were a combination of theory and his practical experience of training officers in Döberitz and Dresden. They formed the basis for officer training and further development of ideas within the General Staff. In 1924, his book *Der Kampfwagen in der heutigen Kriegführung* was published, in which the various types of tanks, tank tactics, anti-tank defence, organizational set-up of tank units and training programmes were described. This book was seen by the German high command as the standard work on warfare with this new weapon, the tank, and fitted seamlessly with the principles of *Führung und Gefecht der Verbundenen Waffen* (leading and fighting with combined arms). This document, written by Hans von Seeckt, translated the principle of *Auftragstaktik* – the German command concept – into principles for the structure and methods of the Reichswehr.

It is interesting to note the great attention that Volckheim paid to anti-tank defence, a result of his experiences in the First World War, when the Allies had a clear superiority in tanks. Volckheim's publications had a great influence on thinking about the future role of the tank and armoured vehicles in the Reichswehr. This was reflected in the articles in the *Militär Wochenblatt*: while in 1923 most articles on mechanized warfare were still written by foreign authors, in 1926 the majority were by Germans. Contributions came from officers of the infantry, artillery, transport units and cavalry, some of whom would achieve a high profile in the years that followed as important thinkers in the field of mobile warfare.

One of these authors was Fritz Heigl, a former Austrian captain and an engineer in the Austrian car industry. He was the best-known analyst in the field of foreign tanks and tank and vehicle technology. From the early 1920s until his death in 1930, he was one of the most prominent authors in the *Militär Wochenblatt*. He also wrote several books on tank technology, including *Taschenbuch der Tanks*, which were distributed within the Reichswehr as approved publications. In addition to Heigl and Volckheim, Lieutenant Wilhelm Brandt wrote regular articles in the *Militär Wochenblatt* on the technical aspects of tanks.

There was also great interest in the motorization and mechanization of the army within the Truppenamt itself. For example, German officers were happy to accept the invitation of the French to attend manoeuvres in 1922 and 1923 as observers. However, they were not impressed because they felt that the French units lacked flexibility and preferred to take up static, defensive positions. They also felt that there were clear weaknesses in the area of communication and cooperation between units. They concluded that horse traction and mechanized units were not compatible and found the use of cavalry as a strategic weapon outdated due to its vulnerability to gas. They were, however, impressed by the French half-tracks and their new armoured equipment, although they considered their own tactical doctrine to be much more effective than that of the French. They had more appreciation for the British tactical doctrine and the degree of motorization of British forces. But the British, too, struggled with the challenges of motorization, and British manoeuvres in 1925 showed once again that the combination of horse-drawn and mechanized units did not work.

To further broaden their horizons, Reichswehr officers were encouraged to take three-month study trips abroad to brush up their languages and gather information. On return home, they had to write a full report on their host country. The United States was, from 1923 onwards, the most popular destination for these study trips, because of the language and the friendly attitude of the Americans. In the US, officers were allowed to travel freely, visit military installations and even inspect secret formations. This meant they could collect information about the motorization and mechanization of the American Army, American vehicles, tanks and artillery tractors, including photos, reports of manoeuvres to which they were invited as observers and details of the structure of units and training.

From 1924 onwards, von Seeckt also focused his attention on the future position of the tank in the Reichswehr. Too weak to take on its enemies, the Reichswehr was motivated to do the things that other armies left undone: it was forced to exploit every advantage and give every innovation a chance. Among many other measures, von Seeckt ensured that tanks would play a role in war games and manoeuvres so that men would get a feel for the possibilities of the tank, learn to work with it and become proficient in anti-tank defence. For example, French Renault tanks and British Mark V tanks were recreated during manoeuvres in 1924. Volkheim's *Der Kampfwagen in der heutigen Kriegführung* and his articles, combined with those of Heigl, resulted in the production of practical manuals and instructions up to company level. It was agreed that all issues in the field of armoured warfare training had to be submitted to the Inspektion der Kraftfahrtruppen, the department where Guderian worked. Guderian evaluated these manoeuvres for the Kraftfahrtruppen, just as other specialist officers did for their respective units. In this way he acquired a thorough knowledge of the possibilities and drawbacks of the new weapon.

It was certainly not the case that Guderian was the only person who shaped the German tank weapon and its use. In his autobiography he exaggerated his own role and, consciously or unconsciously, marginalized that of others: 'Since nobody else busied himself with this material,' he wrote, 'I was soon by way of being an expert.' However, in the 1920s Guderian was only one of many German officers who were involved in research into the potential of the tank and the motorization of the army. He was certainly not, as he himself said, 'a one-eyed man in the country of the blind'. We will return to this later.

Armed with the knowledge of Volckheim, Heigl and others, Guderian was responsible in 1924 for a whole series of manoeuvres with armoured vehicles and dummy tanks, both on paper and in practice. The goal of these manoeuvres was to investigate the possibility of using tanks and cavalry in combination for reconnaissance purposes. The only real vehicles available were four-wheel-drive armoured cars, clumsy machines that could only be used on the road. Guderian's suggestion that motorized supply units could also be used for transporting troops was ridiculed.

After his involvement with armoured vehicles, on 1 October 1924 Guderian was appointed to the staff of the 2nd Division in Stettin, where he had the specific task of training future staff officers in tactics and military history. In history, he focused on Napoleon's campaign of 1806, which resulted in the great German defeat at Jena, as well as on the tactics of the German and French cavalry in the autumn of 1914. Guderian considered the 1806 campaign a good example of a war of movement. His lectures appealed to his audience because of his enthusiasm, and he won many over to his progressive ideas, in which armour more than any other weapon would provide the necessary dynamism on the battlefield. The subjects of his lessons changed over the years from conventional military history to mechanized warfare. As we shall see, Guderian played no role in the development of the first prototypes of German tanks, or in the testing and training centre at Kazan in the Soviet Union. During this fruitful period in Stettin, Guderian's theories on the tactical and operational dimensions of mobile warfare matured.

His ideas did not go unnoticed, and on 1 October 1927, after being promoted to major, Guderian was transferred to the Truppenamt, Heeres-Transport department. His function was newly created and concerned the transport of troops by truck. It soon became clear what this entailed. The French had been very successful in this field in the First World War, for example at Verdun, but their movements had always taken place behind a static front line. In such a situation it is not necessary to transport a complete division with its artillery etc. at the same time. But in a war of movement, all the materiel and equipment of a division must be able to be transported at the same time: the number of trucks required for this task would be unimaginably large. There were heated discussions on whether this would ever be possible, and there were more sceptics than believers in this respect. Due to the small size of the Reichswehr and its limited resources, a more conservative strategy with regard to the deployment of men and equipment was more obviously suitable than the radical and as yet untested theory of a number of junior officers with an enthusiasm for mechanized warfare.

Guderian was also asked in 1928 to teach tank tactics at the Kraftfahrlehrstab in Berlin. This presented an interesting problem: Guderian had never seen the inside of a tank, let alone driven one. Thanks to the abundant

Guderian visiting the Swedish panzer troops in August 1929.

literature available by then, it proved to be less difficult than Guderian had initially thought to get a clear picture of the current state of affairs in the field of tanks and tank tactics. In the summer of 1929, Guderian finally got the chance to drive a tank, when he was sent to Sweden for four weeks to see the latest German tank design, the LK II, in action. The LK II was a German tank design that had been sold to Sweden after the First World War (see below).

In this year he also formulated his ideas about the future position of the tank in the army. In line with the prevailing doctrine from *Führung und Gefecht der Verbundenen Waffen*, the previously mentioned exercises in France and Great Britain and his own observations, Guderian had become convinced that tanks without supporting weapons would never achieve their maximum effectiveness. This meant that these supporting weapons – infantry, engineers, artillery – would have to be able to keep up with the tanks in terms of speed and ability to cope with terrain. In the '*Verbundenen Waffen*' unit that was created in this way, the tanks had the leading role, the other weapons were subordinate. Guderian rejected the inclusion of tanks in infantry divisions: this would greatly reduce the efficiency of the tank.

During summer manoeuvres in 1929, an exercise was carried out under Guderian's leadership with such an imaginary unit. This exercise was, according to him, a success, but the Inspektor des Heeres-Transport, General Otto von Stülpnagel, forbade exercises both on paper and in practice with tank units larger than a regiment. In his opinion, the idea of an armoured division was a utopian pipedream.

Having discussed the development of tank doctrine, we will now consider the technical development of the tank itself.

Chapter 3

The Development of the Tank in the 1920s

The development of tank doctrine cannot be seen in isolation from the technical development of the tank itself in the interwar period. The Treaty of Versailles prohibited Germany from producing or importing armoured cars, tanks and other 'similar war products' (Article 169). In fact, the import of weapons, ammunition and war materiel of any type was strictly prohibited (Article 170), as was the production of war gases (Article 172). Possession of anti-aircraft guns (Article 169) and military aircraft (Article 198) was prohibited, and apart from a few pieces of artillery in various fortifications, the German Army was not allowed to possess more than 204 77mm and 84 105mm guns. For heavy machine guns the number was 792, for light machine guns, 1,134 and for mortars, 252. The Interallied Military Control Commission, consisting of 337 officers and 654 men, was stationed in Germany to monitor the implementation of the Treaty. It would remain there until 1927.

The German government would continually attempt to renegotiate the restrictions on the number of weapons, and in this it had some success. Under the Boulogne Agreement of 1920, Germany,

Thin sheet-steel armoured cars were the forerunners of the German armoured forces' equipment.

partly in view of internal unrest, was allowed to purchase 150 armoured cars for the police, and the Reichswehr was given 105 armoured personnel carriers, which were in fact no more than armoured trucks. The Paris Agreement of 1926 further relaxed the strict restrictions on the production of military aircraft by German industry.

The line taken by the German government was that it was unacceptable for Germany to have an army that could only serve as border guards or an internal security force. In line with this, the Reichswehr tried to circumvent the restrictions in various areas from the day the Treaty came into force and secretly developed a completely new range of modern weapons. The government was aware of the secret rearmament programme and supported it by camouflaging military items in the budget as civilian projects.

Moderate and democratic politicians such as President Friedrich Ebert and Foreign Minister Gustav Streseman unequivocally supported the various secret Reichswehr projects. The German arms industry managed to circumvent the provisions of the Treaty of Versailles in various ways. One of the most effective was the establishment of foreign subsidiaries and the transfer of military production abroad. Krupp, for example, took over the Swedish company Bofors in 1921 and sent German engineers and technicians to Sweden to develop various weapon systems for the Reichswehr. In the Netherlands, Siderius A.G. housed Krupp's artillery and shipbuilding expertise. Rheinmetall took over Solothurn A.G., a watch manufacturer in Switzerland, and switched it to machine gun production. The largest project, however, took the form of cooperation with the Soviet Union in the period 1921 to 1933.

In 1920, a special mission was sent by von Seeckt to the USSR to jointly produce weapons and to set up testing and training centres for tanks and planes. The treaty concluded between Germany and the Soviet Union formed the basis for a completely new generation of aircraft, armoured vehicles and war gases.

The Versailles Treaty's restrictions had advantages and disadvantages for the development of new weapons. The disadvantages were numerous: all new weapons had to be developed abroad and in secrecy, and the funds had to be hidden in various parts of the national budget. For example, the tank programme in the 1920s progressed slowly because all components had to be made in scattered, clandestine workshops by designers sworn to secrecy. Because tanks could not be tested openly in Germany, they were subjected to all kinds of tests in the Soviet Union at the joint test and training centre in Kazan. The designers and engineers could not be present to identify any weaknesses and then make rapid adjustments to the design or construction in the factory. But there were also major advantages: Germany, unlike the Allies, was not saddled with a huge stock of outdated weapons. Consequently, it took a long time for the Allies to develop and acquire new equipment; in fact, the Allies were forced to adapt their tactics to their existing, outdated arsenal, while the Reichswehr was able to develop new tactics first and then the weapons that went with them. For example, the French Army had so many Renault light tanks left over at the end of the war that French tank tactics were based on the capabilities of this 1917 tank for more than a decade. The US Army recognized as early as the 1920s that the 75mm field gun was an unsuitable weapon for a future conflict due to its short range, flat trajectory and relatively small projectile. It was thought that 105mm guns would be more appropriate, but development of these did not begin until the late 1920s. Even then, the army leadership felt that the stockpiles of 75mm shells were so large that the 75mm gun should remain in service until further notice. It was not until 1940 that the 105mm gun was introduced in the US Army, but even then the Chief of Ordnance, General C.M. Wesson, wanted to keep the 75mm gun in service because 'it was an excellent weapon' and 'France had not yet taken it out of service.' The same applied to tanks: the US Army had developed a number of excellent prototypes in the early 1920s, notably those of J. Walter Christie, but they had enough Renault FT 16 tanks left over from the First World War and did not consider it necessary to replace them. The US Army did not acquire any new tank models until the mid-1930s.

In contrast, the German Army developed a whole range of new weapons in the 1920s which were ready for mass production as soon as permission for full rearmament was granted. This was a strange side-effect of the Treaty of Versailles: the restrictions forced the German Army to reinvent itself and prepare for a new type of warfare, while the French and US armies saw no need for innovation and continued to train with outdated weapons and tactics. The end result was that the French Army's arsenal in 1940 consisted partly of 1918 weapons, while the German Army's weapons and tactics were up-to-date. Von Seeckt was right in his view that a small army with superior weapons would defeat a large army with outdated ones.

All in all, in terms of modern weapons, Germany was comparable to other countries at the time; it did not have a clear lead, but nor had the Treaty of Versailles caused it to fall behind. There were, however, important differences. The British and American armies had their own design offices and produced many of their weapons themselves, without using civilian manufacturers. This system worked well: for example, the American Army produced the excellent 105mm howitzer and the M-1 Grant rifle in the period between the two world wars. The Reichswehr, on the other hand, held extensive competitions for every new type of weapon and vehicle. For example, Krupp and Rheinmetall were involved in a fierce battle in the 1920s over almost every piece of equipment, which led to both companies keeping excellent design teams at work. This competition, together with German craftsmanship, was the main reason for the high quality of German weapons in the interwar period. In addition, von Seeckt demanded in 1924 that in the future seminars would be organized every two months for all those involved in the field of weapons technology. The seminars covered subjects such as armoured vehicles, tank technology, motorization, transport vehicles, war gases, modern foreign rifles, developments in the field of American artillery, etc. Through all kinds of such initiatives, a considerable number of the 4,000 officers of the Reichswehr were involved in the research and development of weapons. The Charlottenburg Technische Hochschule served as a cover for technical research, particularly in the fields of ballistics and explosives.

The Reichswehr also kept a close eye on developments abroad, particularly in the United States, France, and Britain. The knowledge gathered here was collected in a series of booklets with the discreet title *Technische Mitteillungen* (technical reports). German officers made targeted visits to the United States in the mid-1920s and returned with a wealth of technical data on the development of explosives and ammunition, artillery and vehicle technology. Much attention was also paid to the effectiveness of French and British equipment during manoeuvres, such as those of 1922 and 1923 in France and the British exercises of 1924.

Motorized armoured vehicles gained increasing recognition as a weapon of the future in the 1920s.

The articles in the *Militär Wochenblatt*, in which Guderian also regularly published, often contained technical descriptions of foreign vehicles, aircraft or guns. Technical ideas were adopted when appropriate, particularly in the field of tracked vehicles and aircraft, but there was seldom any direct copying of foreign designs. The only foreign design adopted without modification was the 7.5cm Gebirgs Kanone (mountain gun) by Skoda.

All these activities made the Reichswehr probably the best informed army in the world on developments within other armies, both technically and in the field of tactics and doctrine. In addition, the new weapon systems had to fit into the larger framework of *Führung und Gefecht der Verbundenen Waffen*; new weapons had to be mobile and have high firepower. In line with this, no more investment was made in weapons intended for fortifications or in materiel for trench warfare.

Armoured vehicles

In the First World War, German car manufacturers had shown that they were capable of developing high-quality tanks. After the war, the Germans sold the components of their most advanced tank, the LK II, to the Swedish

Army. The designer of this tank, Josef Vollmer, also went to Sweden to make a modified version of the tank. The Swedish LK II, called Strv M/21, was equipped with machine guns instead of cannons and entered service with the Swedish Army in 1920. In Germany itself, the revolutionary uprisings in 1919 and 1920 had created a need for armoured cars. Ehrardt GmbH, which had produced such vehicles during the war, built twenty armoured cars in 1919 based on their 1917 model. The Krupp-Daimler chassis formed the basis for another twenty armoured cars built by Daimler. In total, ninety-four armoured cars had been built by January 1920, in violation of the Treaty of Versailles. We have already seen that the Boulogne Agreement eventually allowed Germany to own 150 armoured cars, each armed with two machine guns. In 1921, another eighty-five new armoured cars were built, all of which featured a steering wheel on both sides; this dual steering arrangement would become standard in most German armoured cars in the 1930s and 1940s. The Boulogne Agreement also permitted the construction of 150 armoured personnel carriers, which were designated SdKfz 3 (Sonder Kraft Fahrzeug) as training and test vehicles in the 1920s.

From 1924 to 1938, Oswald Lutz played a central role in the development of armoured vehicles. In 1924, Lieutenant Colonel Lutz was appointed departmental head in the Heereswaffenamt (Army Ordnance Office) to coordinate the development of armoured vehicles. Lutz was the right man for this position: not only did he possess the necessary technical knowledge, but his entire military career had been devoted to transport and supply units, finally as the person responsible for the motorized transport of the Sixth Army. Newly appointed, he noted that no new designs had been drawn up for armoured cars since 1921, and his department was put to work to map out possible developments. This resulted in a large number of programmes in the field of armoured vehicles, namely the development of a 'heavy tractor' in 1925, a 'light tractor' in 1927 and eight-wheel reconnaissance vehicles and half-tracks in 1928.

The 'heavy tractor'

In May 1925, Krupp, Daimler and Rheinmetall were ordered by the Reichswehr to build two tank prototypes each, based on the following specifications: weight 16 tons, armour 14mm, top speed 40kph, an engine developing between 260 and 280hp, a 75mm gun, with a machine gun in the turret and two other machine guns, one of which was aimed from the turret at the rear deck. This last requirement was prompted by the fact that this tank had to force breakthroughs and therefore needed firepower in all directions. The tank had to be able to bridge a ditch 2m wide and negotiate an obstacle 1m high. The crew would consist of six men, and the role of the tank was to support the infantry. The three companies succeeded between 1925 and 1929 in producing tanks which were modern in every respect, and there was a great deal of similarity between them. They were comparable to the British medium tanks of that period, although they had heavier guns (75mm instead of the 3-pounder, 47mm guns of the British) and a heavier engine (a 250hp BMW engine instead of the 180hp engine in the British tanks). In steering, gearbox and other technical details there were differences between the German tanks, but all were amphibious with a propeller at the rear, an idea borrowed from the amphibious tank that Christie had developed for the US Marine Corps in the 1920s. In practice, however, this adaptation proved impractical, and the idea of an amphibious tank was eventually abandoned.

The rear-firing machine gun in the turret proved unnecessary during testing, but the three companies involved and the Reichswehr had learned a lot from these 'heavy tractors'. Moreover, the companies had combined their design teams so that they could gain valuable experience, which would prove its worth in the future. For example, Professor Ferdinand Porsche, the leader of the Daimler design team, would become Germany's most important tank designer.

Tank models become more advanced: in the late 1920s, the body of a tank was mounted on the chassis of a car, as this photo clearly shows.

A wonderful mix of vehicles from the 4th Division at Naumburg in 1927 is shown here, including wooden tanks on a truck, various guns and a soldier on a bicycle.

The 'light tractor'

After the success of the 'heavy tractor', the Reichswehr decided to order the three companies involved to each produce two different prototypes of a 'light tractor'. Daimler ultimately did not participate in the design competition, so that a total of four tanks were available for testing. The development time was now only eighteen months, thanks to the experience of developing the 'heavy tractor' and the departure in 1927 of the Interallied Military Control Commission, which created room for independent weapons development. The design that the Reichswehr had in mind was to serve as a light tank and supply vehicle. The armament had to consist of a semi-automatic 37mm gun and a machine gun. The average speed had to be 25–30kph on the road, with a range of 150km, or 20km off-road. The armour had to be able to withstand 13mm ammunition. The maximum permissible weight was 7.5 tons, and the tank had to be able to bridge a ditch 1.5m wide. The role of this tank was to eliminate enemy tanks with its high muzzle velocity 37mm gun.

Again, the companies were able to deliver state-of-the-art vehicles, but Rheinmetall in particular produced two prototypes that represented a clear leap forward in technology, particularly in terms of the drive train.

The Rheinmetall 'light tractor' of 1928 had many similarities to light tanks in other countries: the engine was mounted in the front and the gun turret at the rear. However, German tanks were more heavily armoured and armed with a cannon and a machine gun instead of just machine guns like the British Vickers tanks.

Three of the four tanks had traditional small road wheels, but one of the Rheinmetall tanks was equipped with four large road wheels and the Christie suspension system. This proved to be a robust design with excellent off-road characteristics. Again, the tanks with their front-mounted engine and rear-mounted turret resembled the British light tanks, but they were more heavily

Another design by Rheinmetall laid the foundation for the design of almost all armoured personnel carriers in the years that followed: the Christie suspension system was combined with separate front wheels for steering.

armed and weighed almost twice as much. The Clectrac drive system used, on the other hand, was derived from one developed by Renault. There was also another not insignificant difference: the German tanks were welded, while the British used rivets.

The wheeled/tracked concept

Krupp developed a third tank design at its own expense in the 1920s. The immediate reason was Christie's designs, which made it possible for a tank to drive at high speed both on- and off-road. A combination of the speed of an armoured car on the road and the cross-country ability of a tank, the so-called 'wheeled/tracked tank' appealed greatly to German tank experts, including Lutz. The concept of a wheeled/tracked tank was popular at the time, and the French had already developed a prototype in the early 1920s. The wheeled/tracked tank was mainly used for reconnaissance purposes and was a logical addition to the two previous tank types. Lutz strongly encouraged the development of such a vehicle, and eventually six examples of the prototype were completed in 1928. The vehicle designed by O. Merker of the Krupp branch Essen was complicated and had four large car wheels and a track system. When off-road, the vehicle used the tracks, while on the road the wheels came down and the track system was pulled up. Three vehicles were equipped with a 50hp Benz engine and the other three with a 70hp NAG engine. The top speed was 46kph on the road and 23kph across country. Lutz would have liked to see the tank able to switch from tracks to wheels and vice versa in a minute and without the driver having to leave the vehicle. However, this was too much to ask of the technology of the 1920s. The vehicle proved to be difficult to control, and in the end the whole project was dropped. At least, that's how it seemed. When the Reichswehr was no longer interested, Merker and his design team were transferred to Landswerk in Sweden, part of the Krupp concern. Here, the wheeled/tracked tank was further developed and put into production in 1930 under the name Landswerk Model 30. The tank sold surprisingly well on the international market, and Krupp earned back its earlier investment in the project. The Reichswehr itself would eventually purchase several examples of the vehicle from the Austrian company Sauer in 1939. They

Another interesting concept to combine high speed on the road with good off-road capability was the Krupp design for a wheeled/tracked vehicle: this is the Austrian version of this design in the early 1930s.

put them into service under the name SdKfz 254, but ultimately they did not meet expectations. Nevertheless, the wheeled/tracked tank also shows that the Reichswehr was well informed about international developments and based its own designs on them.

Reconnaissance vehicles

Based on experience with the wheeled/tracked tank, Lutz and his staff of the Inspektion der Kraftfahrtruppen decided to shift their attention to the possibility of using six- and eight-wheeled armoured vehicles and half-tracks for reconnaissance purposes. This was incorporated into the existing programme for the development of armoured cars. In 1927, Büssing, Daimler and Magirus were commissioned to develop armoured cars that had to be able to reach a top speed of 65kph, be able to cross a 1.5m ditch and negotiate a 33° gradient. The cars had to have a steering system on both sides. The weight was to be a maximum of 7.5 tons. The designs of the three companies were outstandingly original and surpassed those of other countries at the time. The Porsche design team of Daimler produced a fast eight-wheeled vehicle with a low silhouette, a 100hp engine, four-wheel drive and 13.5mm of armour. The Magirus vehicle was similar in many respects to Daimler's. Büssing produced a ten-wheeled vehicle with good speed and good off-road capability, but the braking of ten wheels at high speed was dangerous. This prototype was eventually discarded. The Reichswehr placed the necessary orders with Daimler and Magirus, and Rheinmetall supplied the gun turrets with a 37mm gun and a machine gun. This programme convinced the Reichswehr of the usefulness of eight-wheel armoured vehicles, and the Daimler and Magirus designs were the direct predecessors of the famous SdKfz 231 armoured reconnaissance vehicles of the Second World War.

Half-tracks

The development of half-tracks began in 1930 as part of a broader programme to increase the motorization of the Reichswehr. Lutz and his staff had been toying with the idea of using half-tracks to transport troops since 1928. Half-tracks, in which the rear axle of a truck was replaced by two or more rollers with a track around them, were a proven technology from the First World War. The Allies were the first to use them, but in 1918 the German Army introduced the Benz-Bräuer half-track truck, an adaptation of an existing truck type. After the war, in the mid-1920s, Dürkopp GmbH designed a vehicle for heavy agricultural work with its own drive. Maffei GmbH also produced a civilian half-track vehicle based on French licences from 1927 that could carry eight people or 1,000kg of cargo. The Reichswehr and the Austrian Army soon bought several examples of this vehicle, and Maffei would continue to develop half-track vehicles for the Reichswehr and later the Wehrmacht.

The leading role of Lutz and his staff is clearly evident in this sketch of the development of armoured vehicles for the Reichswehr. Lutz was appointed chief of staff of the Inspectorate of the Kraftfahrtruppe in 1928 and would eventually see his career crowned in 1935 by his appointment as the first commander of the Panzertruppen. He can rightly be considered one of the central figures in the mechanization of the Reichswehr after the First World War.

Chapter 4

The Tank Weapon Begins to Take Shape

Guderian would join Lutz at the end of the 1920s to give German armour its definitive form. Lutz, now a colonel, was appointed chief of staff of the Inspektion der Kraftfahrtruppen under Inspektor-General Colonel Alfred von Vollard-Bockelberg in 1928 after developing armoured vehicles at the Heereswaffenamt. Lutz intended to have the seven transport battalions of the Kraftfahrtruppen under his command train as motorized units equipped with dummy tanks, dummy anti-tank guns, the existing armoured vehicles and the first tank prototypes.

In doing so, he clearly gave shape to the ambition of the Kraftfahrtruppen to play a central role in the field of mechanized warfare in the future. Lutz approached Guderian in the autumn of 1929 to ask whether he would like to command one of these motorized battalions. Guderian gladly agreed and on 1 February 1930 he became commander of the 3rd Kraftfahrtabteilung of the 3rd Division in Berlin-Lankwitz. This unit was then structured as follows: Companies 1 and 4 were stationed with the staff in Berlin-Lankwitz, Company 2 was in the military training area of Döberitz and Company 3 was stationed on the Neisse. Lutz had his own ideas, and after Guderian's appointment the unit was re-equipped: Company 1 was given armoured reconnaissance vehicles, Company 2 was given dummy tanks, Company 3 was organized into an anti-tank unit with dummy guns and Company 4 was given motorcycles. In fact, only Company 4 had deployable vehicles, namely the motorcycles: it was decided to spare Company 1's armoured reconnaissance vehicles from the Versailles period as much as possible. However, the officers and men were enthusiastic about their changed role: instead of being in supply, they were now a combat unit! Thanks to cross-training in the Reichswehr, they had little difficulty in adapting to their new role. Further up in the hierarchy, the situation was different: Inspektor-General Colonel Alfred von Vollard Bockelberg had little faith in the new unit and forbade it to undertake joint exercises with other units in the region. Even when the 3rd Division did exercise with other units, their role had to remain limited and they were not allowed to operate in formations larger than a platoon. However, the commander of the 3rd Division, General Joachim von Stülpnagel, was positive about the experiments of Lutz and Guderian and gave them space in his division. Guderian's battalion was not the only one that was equipped in this way; the other six battalions of the Kraftfahrtruppen in the other Wehrkreisen had a similar structure and encountered the same resistance to a greater or lesser extent.

The situation changed in favour of the progressive forces, however, when Lutz succeeded von Vollard-Bockelberg as Inspektor-General der Kraftfahrtruppen in 1931. Guderian became his chief of staff in the autumn of the same year, and in the following years both men began to give more substance and shape to their ideas about armoured forces. Lutz, with his communication skills and political nous, would operate more in the background in the coming years, while the more persistent and forceful Guderian confronted sceptics and outspoken opponents. This may have given the impression that Guderian was the main player, but in reality it was Lutz, given his seniority, who was able slowly but surely to expand the influence of the Kraftfahrtruppen in the coming years. In any case, their joint goals were, on the one hand, to give the German Army its own tank force, and on the other, to convince the infantry and cavalry of the decisive role that tanks could play at the operational level. It would cost them considerable effort to convince the German high command that the motorized supply units, because that is what they still were, were capable of developing and putting into practice such far-reaching ideas in the field of tactical and operational doctrine. The cavalry and infantry were traditionally still the most important parts of the army, with the infantry being the 'queen of the battlefield'. Since the Reichswehr was not allowed to own tanks at all, men had to use their imagination; no one had ever seen these new weapons in action, and the imitation tanks that were used during manoeuvres aroused only disdain in most observers.

In line with this, the other army units saw the tank at most as a weapon that could support the infantry; certainly not as a new independent part of the army and their future competitor. Such views were in fact to be found

The Inspektorat der Kraftfahrtruppen in 1933. Seated L to R: Walter Nehring, Oswald Lutz, Heinz Guderian. Standing right: von Hünersdorff.

in all countries that were concerned with the question of what role the tank could play in a future conflict. It should be noted, too, that tank development and production was taking up a disproportionately large share of often limited military budgets.

Not unexpectedly, a major opponent of Lutz's ideas was the Inspektorat der Kavallerie. In order to clarify the different positions, Lutz asked his counterpart in the Inspektorat der Kavallerie, General von Hirschberg, whether the cavalry saw its future role as a tactical reconnaissance arm or in carrying out strategic reconnaissance and deep penetration behind enemy lines. Von Hirschberg replied that the latter would be the primary role of the cavalry and that the task of tactical reconnaissance would gladly be transferred to the Kraftfahrtruppen. One of the reasons why von Hirschberg so easily relinquished this role was that he could not imagine how Lutz could use the armoured vehicles available at that time – only suitable for use on paved roads – for reconnaissance purposes. This gave Lutz the chance to set up armoured reconnaissance units within the Kraftfahrtruppen and to train them for a reconnaissance role. However, Lutz did not get away with it that easily and von Hirschberg's successor, General Knochenhauer, attempted – in vain – to reclaim the tactical reconnaissance role for the cavalry. Lutz had therefore managed to give the Kraftfahrtruppen an excellent start in discussions about the future role of the tank.

In the meantime, the Inspektion der Kraftfahrtruppen was thinking hard about what future tanks should look like. The prototypes built in the 1920s already represented a great step forward from the tanks of the First World War, but they did not yet meet the tactical requirements of their new role. The chief weak point in the designs was the position of the commander. He sat in the hull next to the driver, instead of in the turret, and thus had a limited field of vision; the lack of radio equipment was another drawback. Von Vollard-Bockelberg and Lutz had already expressed their preference for two new types of tanks at the end of the 1920s: a light tank with an armour-piercing gun and two machine guns, one in the turret and one in the hull, and a medium tank with a heavier gun for high explosive shells as well as two machine guns, similarly positioned. The armoured division they had in mind would be made up of three companies of light tanks and one of medium tanks. The medium tanks would support the light tanks in the breakthrough and then serve to clear out any remaining pockets of resistance.

Lutz and Guderian continued to work along the lines set by von Vollard-Bockelberg and advocated to the Heereswaffenamt the development of a 50mm gun with armour-piercing ammunition, especially in view of what was expected to be the increasingly heavy armour of foreign tanks. Since German industry was not yet able to produce such a gun, Lutz and Guderian had to make do with the 37mm gun used by the infantry. It was decided, however, to develop a 75mm gun for the medium tank. The following additional specifications were formulated for both types of tank: the weight of the medium tank was not to exceed 24 tons – the maximum load-bearing capacity of German bridges at the time – its top speed had to be 40kph and the crew had to consist of five: gunner, loader, commander, who had to sit above the gunner in the turret with a special command cupola from which he could view the entire area, driver and radio operator. In addition, the crew members had to be equipped with throat microphones for mutual communication. The tanks were to be equipped with radio communication equipment so that they could maintain contact with each other and the command centre during a rapid advance behind enemy lines. This important development was undoubtedly Guderian's contribution, given his background in the field of communications in the First World War. In addition, radio communication had been an important research subject within the Reichswehr for many years and it fitted seamlessly with the idea of motorization. Equipping all tanks with radio equipment would prove to be a huge undertaking, but it would ultimately contribute significantly to the effectiveness of German armour.

Lutz and Guderian realized that it would take years before this new generation of tanks would be available. In the meantime, a training tank was needed. The British Carden-Loyd chassis was chosen, and the first units were purchased in 1932. The Carden-Loyd chassis was in fact intended as the platform for a 20mm anti-aircraft gun and only offered space for a turret with machine guns. But despite this disadvantage it was considered an interesting interim solution, and formed the basis for a design by Krupp under the name PzKpfw 1 in 1934. In the meantime, it had become clear that the development of the light and medium tank would take much longer than expected, and Lutz decided on a second interim solution, the PzKpfw II with a 20mm gun and a machine gun. The latter was developed and produced by MAN. No one could have imagined that these two tanks, which only served originally as an interim solution, would ever play an important role in future military operations.

During the summer, Lutz organized manoeuvres for the first time with a combination of reinforced infantry regiments and tank battalions, the latter,

The Krupp chassis, which became the basis for the PzKpfw I, was tested in the Soviet Union in the early 1930s.

of course, still equipped with dummy tanks. During these manoeuvres, the first generation of armoured reconnaissance vehicles, which had been built on the basis of specifications for the Kraftfahrtruppen, could also be deployed. These were vehicles with a steel body on the chassis of a six-wheeled truck. As Guderian notes in his memoirs, that year, the schoolchildren who were used to sticking pencils through the canvas walls of the dummy tanks were just as disappointed as the infantrymen. The new vehicles were not to be trifled with, and the bayonet was no longer an effective weapon against them. During the manoeuvres, the combination of motorized and armoured units was tested. Despite the expected criticism from the cavalry, interest in the new weapon began to grow within this branch of the armed forces, and for the first time a number of young cavalrymen considered making the transition to the Kraftfahrtruppen: they realized that the traditional functions of the cavalry could better be carried out in the future with these new weapons. New developments would further strengthen the position of the Kraftfahrtruppen.

The coming of Hitler

In 1933 Hitler became Chancellor and revolutionized the domestic and foreign policy of Germany. Guderian saw Hitler for the first time at the opening of the Berlin Automobile Exhibition in 1933. His speech was different to those of his predecessors: Hitler opted for the development of the car and abolished the tax on private vehicles. He also presented his plans for a nationwide system of autobahns and the development and mass production of a car for the people, the Volkswagen.

Unsurprisingly, the development of an armoured force gained momentum in 1933 and 1934. On 1 November 1933, Lutz established the first tank training company in Zossen near Berlin. A series of experiments with mock-ups strengthened Lutz and Guderian in their conviction that tanks could only be used to their full potential if they were considered as a central weapon system and accompanied by fully mechanized support units. However, there were still great practical problems; for years, German industry had been prohibited from producing military equipment by the Treaty of Versailles. As a result, it lacked not only skilled personnel but also the necessary machinery.

Hitler witnesses for the first time the possibilities of the tank, Kummersdorf, 1934. On the left, standing by the car and saluting, is Guderian. Hitler in a light coat is on the right.

In particular, the production of sufficiently hardened steel, optical aiming equipment and radio equipment caused many initial problems. Once these were overcome, German industry proved able to deliver advanced products of top quality which would more than compensate for the inherent weaknesses of German tanks in the years to come.

When Freiherr von Fritsch was appointed as Chef der Heeresleitung (commander-in-chief of the army) in the autumn of 1933, all signals seemed to be green for Lutz and Guderian. Although von Fritsch did not have the necessary technical knowledge, he had always shown great interest in the development of motorized units and had also studied the potential of armour for some time. Ludwig Beck, the newly appointed Generalstabchef des Heeres (Chief of the General Staff), took a completely different position: he was not an advocate of independent armoured units and saw the role of the tank primarily as support for the infantry.

In the meantime, Hitler had shown great interest in the mechanization of the Reichswehr. Lutz and Guderian were given their first chance to demonstrate the capabilities of armoured units in early 1934 at a meeting organized by the Heereswaffenamt for Hitler at Kummersdorf on the subject of recent weapons developments. The Kraftfahrtruppen were allotted half an hour and gave a demonstration with a motorcycle platoon, an anti-tank platoon, a platoon of PzKpfw Is in experimental form and a platoon of light and one of heavy armoured reconnaissance vehicles. Hitler was impressed by the speed and precision of the units and is said to have declared, 'That's what I need, that's what I want to have.' What exactly he meant by this is unclear; perhaps he saw the armoured troops as a good propaganda tool to demonstrate the modernity of the Nazi regime, or as a deterrent to neighbouring countries that were following developments in Germany with a wary eye.

We must realize that given the political and military reality at that time, the only role envisaged for the Reichswehr, including tanks of course, was a purely defensive one, namely protecting Germany against possible aggression by France, Czechoslovakia and Poland. For example, Lutz and Guderian saw mobile units, including tanks, as disrupting a possible enemy invasion by attacking their flanks and rear. There was some thought about breaking through enemy lines, but only to strengthen German defences. In that respect,

A first parade with the PzKpfw I in Potsdam in 1935. Oswald Lutz stands next to the panzer, while Heinz Guderian is in front of it.

Lutz and Guderian followed the prevailing view of national defence and military doctrine as formulated in the 1920s.

Developments now gained pace: in the spring of 1934, a Kommando der Kraftfahrkampftruppen was established, with Lutz, now a general, at its head and Guderian as its chief of staff. Lutz combined his role with that of Inspektor der Kraftfahrtruppen to ensure a smooth transition. In August 1934, an embryonic armoured division, composed of units from the Kraftfahrtruppen, demonstrated the potential of mechanized units.

German rearmament

Crucial to the further development of armour was the coming of German rearmament. In March 1935 the Treaty of Versailles was officially terminated, and on 16 March general conscription, initially for twelve months, later for two years, was reintroduced. The seven divisions of the 100,000-man army would be expanded to seventeen, and the Kriegsakademie would be reopened. Importantly, the expanded army, renamed the Wehrmacht, included three panzer divisions. This meant a huge investment in a weapon system that had not yet had time to fully prove its worth. On 15 October the three divisions were officially established, and Guderian was given command of one of them: the 2nd Panzer Division in Würzberg.

However, there were a good many officers in the Reichswehr who were less enthusiastic about the recent developments, and Chief of Staff Beck was one of them. Beck is often seen as a conservative and one of the great opponents of Lutz and Guderian. But Beck's interests went further than those of Lutz and Guderian: he had to develop the Reichswehr into a balanced fighting force and was resistant to allocating substantial resources to an untried weapon system. That he nevertheless did so indicates that he recognized the importance of the tank in future conflicts and dared to make strategic choices: with 'only' three armoured divisions in the total of ten newly created divisions, not only were approximately 30 per cent of the men assigned to the new armoured units, but also a considerable part of the available budget, because the investment in tanks was greater than in other weapon systems at the time.

The infantry and artillery felt that these developments threatened their position; they demanded that the tank should first successfully demonstrate its usefulness, and they also questioned whether it was the most cost-effective use of scarce resources. If the claims of Lutz and Guderian were built on sand, then the Reichswehr lacked the time and resources to develop alternatives.

Although Guderian in particular continually advocated that all armoured units should be placed within the new armoured divisions, separate from the existing military with their prejudices and traditions, the reality was more complicated. For example, within the cavalry there were many young officers who eagerly embraced the possibilities of mechanization as extremely relevant to their role. The cavalry would ultimately provide many of the officers for the future armoured formations, and many of the successful commanders of armour in the Second World War – Ewald von Kleist, Eberhard von Mackensen and Freiherr Geyr von Schweppenburg – came from the cavalry. Cavalry formations were already among the most motorized and mechanized units within the Reichswehr at the beginning of the 1930s, and the 3rd Kavallerie Division had also played a central role as a partially mechanized unit in the annual manoeuvres of 1934. It is significant that Guderian's colleagues in the other two newly established panzer divisions came from the cavalry: the leading role of the Kraftfahrtruppen described by Guderian in his memoirs was apparently not recognized by everyone.

Nor had the possibilities of mechanization passed the artillery by. As early as the 1920s, the artillery was experimenting with the possibilities of mechanization by placing guns on agricultural tractors. The role that Guderian envisaged for them as mobile support for panzer troops, however, far exceeded the communication and fire control techniques of the time. This did not detract from the fact that mechanization was also high on the agenda for the artillery; for example, they had already experimented with a mechanized battalion during the manoeuvres of 1932. They also argued for a mechanized artillery battalion as part of the newly created panzer divisions. Guderian's attempt to remove these units from the artillery and transfer them to the panzer divisions was understandably fruitless: in the first place, Guderian was 'only' chief of staff of the Kraftfahrkampftruppen, and the artillery had high-level patrons: leading figures in the Wehrmacht in the 1930s such as Keitel, Beck and Halder all had an artillery background.

Two rows of PzKpfw I Ausf. A and B tanks waiting at the Brandenburg Gate to take part in a parade, Berlin, 1936. There are twenty-two vehicles in the picture, almost an entire company. The difference in length of 400mm between Ausf. A and B is clearly visible in the right-hand row of tanks.

When the artillery battalion did not become an integral part of the panzer division, Lutz and Guderian conveniently deleted it from the panzer division's organizational chart.

The PzKpfw I contributed little to reducing the scepticism of the rest of the armed forces: small, weak, lightly armed and armoured and with limited off-road capability, this did not seem to be a weapon able to successfully carry out deep penetration behind enemy lines in the future. But the PzKpfw I and the PzKpfw II were the only tanks that German industry could produce at that time.

Panzer divisions

In the second half of the 1930s, the organizational forms and weapon systems that would play an important role during the Second World War were slowly but surely coming into being. In considering them, we will start with the panzer divisions of Lutz and Guderian. The organizational chart drawn up by Lutz and Guderian in August 1935 assumed a 'heavy' armoured division with no fewer than 481 tanks in addition to infantry in armoured personnel carriers, the design of which did not even exist yet. The division was to consist of four tank battalions, three motorized infantry battalions, a reconnaissance battalion, an artillery regiment, an anti-tank battalion, a communications battalion and an engineer unit. The division was amply supplied with communication equipment, the result of Guderian's experiences as a communications officer. In the First World War, divisional commanders had been forced by the slowness of communications to draw up orders based on information that was several hours old. Since it took at least as long for these orders to reach the units at the front, commanders and their troops were effectively fighting wars in parallel worlds. Guderian wanted to reduce this lead time to a few minutes by introducing radio equipment to all tanks and armoured vehicles, and this became a standard concept in the Reichswehr.

Guderian was very fond of the PzKpfw I because he assumed that mobility would be of great importance in any future conflict, and he rejected criticism of the weak armament and armour of this tank. In the course of 1935, the Reichswehr received the PzKpfw II as a reconnaissance tank and the PzKpfw IV medium tank with a 75mm gun for HE ammunition in the direct

The PzKpfw II was the Wehrmacht's first true battle tank in the late 1930s. Production of the tank began in May 1936, and 1,223 were available at the outbreak of war. Here we see a PzKpfw II during training in 1939/1940.

A PzKpfw III A negotiates an obstacle during a demonstration. In May 1937, only five examples of the PzKpfw III were available, and by the end of 1937 only ten more had been added. The armament consisted of a 37mm gun, two 7.92mm MG 34 machine guns parallel to the gun and a 7.92mm machine gun in the hull.

fire support role. These tanks were lightly armoured and vulnerable to direct hits from the existing anti-tank and artillery guns. The PzKpfw III with its 37mm gun was to fulfil the future role of the tank 'killer'. A not unimportant detail was that the turrets of both tanks were designed in such a way that a gun of a larger calibre could be installed in them easily. The way these 'heavy' armoured divisions were organized evolved through experience and the changing views of the high command, to take on by 1940 the definitive shape of the panzer division.

Mechanization of infantry

While Lutz and especially Guderian wanted to ensure that the financial resources for mechanization would be devoted solely to the armoured divisions, Beck continued with the mechanization of the Wehrmacht in a broader sense. When the infantry's request for motor vehicles to replace their horse-drawn anti-tank guns and artillery was accepted, Guderian called this an unwise decision. However, the mobility and effectiveness of the infantry improved considerably, enabling it to play an important role during Blitzkrieg. In addition, at Beck's initiative, the infantry considered the possibilities of further mechanization. This was logical, because in a future mobile war the infantry would have to be able to keep up with other mechanized units such as the armoured divisions. Eventually, the infantry managed to lightly mechanize three, subsequently four, divisions. These included a small armoured unit that could undertake reconnaissance and deep penetration at an operational level, activities normally carried out by the cavalry. In addition, four infantry divisions were fully motorized. Guderian considered this a waste of resources and at that time still assumed that armoured divisions could act independently on the battlefield, while the key to the success of Blitzkrieg turned out to lie in a combination of armoured units and mechanized units of artillery, infantry and engineers, supplemented by the Luftwaffe in its ground support role. There was a further consideration in the creation of these lightly mechanized infantry divisions: the 'heavy' armoured divisions required so much logistical support that serious doubts existed about their operational mobility. These doubts were confirmed during the Anschluss and manoeuvres in the late 1930s. The lightly mechanized infantry divisions, on the other hand, were seen as more or less self-sufficient units, better able to quickly achieve deep penetration and subsequent exploitation of a breakthrough. They were also seen as a complement to, rather than a competitor of, the existing armoured divisions.

Assault guns

On the same grounds, Guderian criticized the new concept of the assault gun, an artillery howitzer on the modified chassis of a PzKpfw III. This would significantly increase the firepower of the infantry. In his memoirs, Guderian criticized the conservatism of the artillerymen, which led to only five prototypes being available in 1939, when Germany went to war with Poland. But the truth is more nuanced: in 1935 the artillerymen had already proposed adding a fully mechanized field artillery battalion to the panzer divisions that Lutz and Guderian had in mind. When disagreement about the authority over these battalions escalated, Lutz and Guderian simply removed them from the organizational chart, as we have seen. Here too, it was Beck who, in the second half of the 1930s, encouraged the artillerymen to further develop their ideas on the mechanization of artillery in order to achieve a balanced structure of the Wehrmacht, with sufficient offensive capacity for the infantry. While Guderian opposed the use of the PzKpfw III chassis for the assault gun, this Sturmgeschütz (StuG) would prove to be one of the most successful combat vehicles of the Second World War. The possibility of upgrading its armament and armour over the years made it the most produced, most cost-effective and most efficient tank killer of the Second World War. One weak point of the artillery would persist, however: the heavy field artillery remained largely dependent on horse traction, which would prove to be a challenge, particularly on the Eastern Front.

Reconnaissance units

In the second half of the 1930s, reconnaissance units were again incorporated into the cavalry. This was logical in itself, because conducting reconnaissance had been the cavalry's most important task since time immemorial. In addition, the leadership of the Wehrmacht recognized the need to be able to deploy lighter and more mobile mechanized units in addition to the 'heavy' armoured divisions.

While the motorization and mechanization of the Wehrmacht progressed on paper, its practical application proved to be more difficult: Germany simply did not have the industrial capacity in the 1930s to meet the military's expectations. The result was that during the Second World War the level of motorization and mechanization declined continuously, until in 1943 this downward trend was reversed by the efforts of Speer and Guderian. But by then it was already too late to turn the tide of war.

Concentration of mechanized units

One of the most important organizational steps in mechanization was the establishment of the world's first corps-level command for mechanized units in 1938, the XVI Armee-Korps, which included all three panzer divisions. This was combined with corps commands for the lightly mechanized infantry divisions and the motorized infantry divisions under Gruppenkommando IV.

Colonel General von Brauchitsch, who had organized the first mechanized manoeuvres in 1923, became commander of this Gruppenkommando. In line with this, the focus for the deployment of mechanized units shifted from defence to more offensive operations.

Achtung-Panzer!

Guderian, meanwhile, was only marginally involved in all these developments. In Würzberg, he was busy putting together and training his panzer division, which was made up of units with very diverse backgrounds. In the spring of 1936, Hitler re-occupied the Rhineland, an operation for which Guderian's division was kept in reserve. The division was moved to the training area in Münsingen, but in order to avoid increased tension, without its panzer brigade. The occupation was carried out without problems, and the units returned to Würzberg within a few weeks.

In the autumn of 1936, Lutz encouraged Guderian to write a book with the aim of further promoting panzer divisions and the concept of independent, strategically operating mechanized units. An underlying reason may have been that the role of armoured units was coming increasingly under pressure and that, as we have seen, investment in the field of mechanization did not only benefit the armoured divisions. Guderian's book, *Achtung-Panzer!* was based on his lectures and articles, combined with some historical reflections, and relied heavily on the ideas of other tank theorists, in particular Ludwig Ritter von Eimannsberger. The book was published in the spring of 1937 and was a resounding success: with his royalties, Guderian was able to buy his first car. In an article in the October issue of the magazine for regular officers he once again emphasized the need to deploy tanks in a concentrated manner. As he put it, an infantry division with fifty anti-tank guns would be able to withstand an attack with fifty tanks more easily than an attack with two hundred tanks. He then concluded that the idea of providing infantry divisions with tanks meant a return to the British tactics of 1916/1917, which had proved to be a failure. The British had only been successful when tanks were deployed in large numbers, as at Cambrai. Only if an attack was carried out with a sufficient concentration of tanks in width and depth would they be able to breach the enemy's defences, through which the units that followed would be able to advance more quickly thanks to their better mechanization than in 1918. This fighting force should, in Guderian's opinion, consist of panzer divisions, since he himself – as he wrote – no longer believed that other formations, such as infantry, had the fighting power, speed and manoeuvrability to successfully force a breakthrough and exploit the opportunities that would then arise. His message was that three panzer divisions were in fact too few, and that the materiel of the three lightly mechanized infantry divisions should be used to form more panzer divisions. However, Guderian was not in position at that time to influence the policy of Beck and the high command.

In the autumn of 1937, manoeuvres were held involving the 3rd Panzer Division and the 1st Panzer Brigade. The spectators included Hitler, Mussolini, British Field Marshal Sir Cyril Deverell, Italian Marshal Badoglio and a Hungarian military mission. Guderian was one of the referees during these manoeuvres, which left a positive impression on all those present. On the last day of the manoeuvres, a final exercise, this time under the command of Guderian himself, was carried out, in which all the tanks were deployed for the foreign guests. It was an impressive display, despite the fact that the only tank available was the PzKpfw I. A weak point that emerged during these manoeuvres was the shortage of supply and repair facilities, something that would affect Guderian and his 2nd Panzer Division in the spring of 1938, when they were deployed in the Anschluss with Austria.

The year 1938 began with a surprise: Guderian was promoted to lieutenant general on 2 February and appointed commander of the XVI Armee-Korps. He learned of this promotion and appointment from the same newspaper that reported the sacking of a large number of high-ranking officers of the Wehrmacht, including Fritsch and von Blomberg. Lutz, General der Panzertruppen, the man who had played such a leading role in the development of the armoured weapon for almost two decades, had also been dismissed. The reason given for von Blomberg's dismissal was his second marriage. Fritsch was dismissed because he was under judicial investigation. No reason was given for the other dismissals. To top it off, Guderian, like all the other generals, was summoned to attend a meeting chaired by Hitler on 4 February. The dismissals sent shock waves through the officer corps, because all those involved were men of impeccable conduct.

The image and prestige of the Wehrmacht were seriously damaged by this, and the dismissals illustrated the gulf that existed between it and the Nazi leadership. The dismissal of von Blomberg was perhaps justifiable by the spirit of the times, but the trial of Colonel General Freiherr von Fritsch before a military court chaired by Göring yielded no evidence, and he was acquitted. However, he did not return to service, becoming an advisory colonel in an artillery regiment.

After the dismissal of these key figures, Hitler dramatically strengthened his position by appointing himself commander-in-chief of the Wehrmacht and relinquishing the post of Minister of Defence. The newly appointed General Wilhelm Keitel took over a number of ministerial tasks. However, Keitel had no formal powers and called his position 'Chef des Oberkommando der Wehrmacht'. General von Brauchitsch became commander-in-chief of the Heer and Reichenau commander of the Gruppenkommando IV. The wave of dismissals and the resulting reshuffling of functions also offered opportunities for Guderian: both his new job and the appointment of Wilhelm Keitel gave him access to the highest levels in the Wehrmacht's hierarchy, and also increasingly to the circle around Hitler.

The role of Guderian

It may be useful to look back at Guderian's role in the development of German armour between the two world wars. First of all, Guderian undeniably played an important role in the development of German tank doctrine in particular. The central role generally attributed to Guderian was in large part due to his autobiography *Errinerungen eines Soldaten*, which was published in English under the title *Panzer Leader* and was a worldwide best-seller. In the book, Guderian boosted his own role and thereby marginalized other people who were also important to the development of armour. As a result, Guderian became, perhaps after Rommel, the best-known of German generals. Moreover, his name is the one almost exclusively associated by many writers with the development of the German tank force before the Second World War.

The previous chapters have shown, however, that Guderian was certainly not the only man responsible. That would have been impossible; for years, he simply held too lowly a position in the military hierarchy to play a decisive role. Guderian did write five articles for the *Militär Wochenblatt* between 1922 and 1928, but these concerned relatively unimportant topics such as 'French motorized supply units at Verdun' and 'Reconnaissance and protection of motorized units'. His contributions on tactical subjects were relatively limited and focused on 'Cavalry and Armoured Cars' and 'Troop Transport Vehicles and Air Defence'. These were not revolutionary articles but rather reflected conventional opinion at the time. *Achtung-Panzer!*, written in 1936/1937, on the other hand, was a brilliant and original book that can be seen as a crystallization of the ideas of tank theoreticians in the preceding decades. It built in particular on ideas in the book *Der Kampfpanzerkrieg* by the Austrian (and later German) General Ludwig Ritter von Eimannsberger, a publication that enjoyed wide recognition in the Reichswehr.

In his autobiography, Guderian completely ignored all the other officers who played a role in the development of German armour. In comparison, General Walther Nehring gives a detailed list in his book *Die Geschichte der Deutsche Panzerwaffe 1916 bis 1945* of the many dozens of officers who played a role directly or indirectly in various parts of the Reichswehr. These included the Heereswaffenamt, the experiments of the Inspektorat der Kraftfahrtruppen and everyone who was involved in the training and testing centre in Kazan. This, combined with von Seeckt's ideas about manoeuvre warfare and the cooperation between the various weapon systems as shown in *Führung und Gefecht der Verbundenen Waffen*, ultimately provided the technical and theoretical basis for the German armoured force.

In addition, we should not underestimate the role of Oswald Lutz, and his effect on Guderian's career. Lutz played a central role in the field of armoured vehicle development for more than a decade in the 1920s and early 1930s, crowned in 1935 with his appointment as the first commander of the Panzertruppen.

The role of British theoreticians

The role of British theoreticians such as Fuller and Liddell Hart has also been considerably exaggerated thanks to Guderian's book *Panzer Leader*.

Guderian is not to blame for this misconception, but Liddell Hart is. In the first editions, Guderian was initially translated correctly:

> It was principally the books and articles of the Englishmen Fuller, Liddell Hart and Martel, that excited my interest and gave me food for thought. These far-sighted soldiers were even then [1920s] trying to make of the tank something more than just an infantry weapon. They envisaged it in relationship to the growing motorization of our age, and thus they became the pioneers of a new type of warfare on the largest scale.

So far, nothing wrong. But in the subsequent English edition, Liddell Hart as editor committed a clever piece of falsification of history. He added the next paragraph:

> I learned from them (Fuller, Liddell Hart, Martel) the concentration of armour, as employed in the Battle of Cambrai. Further, it was Liddell Hart who emphasized the use of armoured forces for long-range strokes, operations against the opposing army's communications, and also proposed a type of armoured division combining Panzer and Panzer-infantry units. Deeply impressed by these ideas, I tried to develop them in a way practical for our own army. So I owe many suggestions of our further development to Captain Liddell Hart.

Liddell Hart wanted no less than to be recognized as the spiritual father of Blitzkrieg tactics and regularly claimed to be so. Many historians have accepted this claim without question, but closer examination of German books and articles shows that Liddell Hart, in fact a junior officer in the 1920s just like Guderian, was largely unknown in the Reichswehr and that his ideas in no way influenced German doctrine and tactics. Although some of his short articles from the *Daily Telegraph* were published in the *Militär Wochenblatt* in the 1920s, the same was true of many hundreds of French, British, American, Polish and Italian writers. Neither Guderian, von Eimannsberger, Heigl or Volckheim quote Liddell Hart anywhere or show any knowledge of his ideas. Erwin Rommel, for example, first heard of them at the end of 1942.

J.C. Fuller, on the other hand, was well known in Reichswehr circles. His fame was not so much based on his theories, but on his role as Chief of Staff of the British Tank Corps. He had prepared the massive tank attack at Cambrai in 1917 and was the author of Plan 1919, which we will discuss later. Fuller's first book, *Tanks in the Great War*, was particularly appreciated for its useful information on the composition of tank units, the maintenance and supply of tanks, etc., as well as its description of the lessons of the First World War. His second book, *The Reformation of War*, also received a great deal of attention from the Reichswehr, and his memoirs were translated into German. However, the German military studied many and diverse sources, analysed them critically and formulated their own ideas about the potential that tanks and armoured vehicles offered them within the restrictions imposed by the Treaty of Versailles. For example, the organizational structure of tank formations in the 1920s and 1930s was largely derived from Fuller's ideas, but his tank tactics were not taken seriously by the Germans. Volckheim did indeed adopt some of Fuller's ideas in 1924, but he rejected Fuller's enthusiasm for small, fast tanks, the one- and two-man 'tankettes'.

All in all, the development of German armour tactics was based on the command concept of *Auftragstaktik* and *Führung und Gefecht der Verbundenen Waffen*. If there was any degree of influence from British theoreticians, this honour goes to Fuller, alongside many others, and not to Liddell Hart. In the next chapter we will examine how the deployment of mechanized and motorized units fitted into overall German doctrine.

Chapter 5

German Panzer Doctrine

To understand German panzer doctrine in its mature form, we need to go back to the end of the First World War and the command concept that emerged in the years that followed. Panzer doctrine, embedded in this command concept, would prove to be extremely successful in practice.

As we have seen, the German Army faced major challenges after the First World War. The peace treaty of Versailles limited its size to 100,000 men, including 4,000 officers. It was allowed to consist of seven infantry and three cavalry divisions, and the weapons permitted to these divisions were described in detail. The Reichswehr was only allowed light weapons and field guns; heavy field guns, tanks and aircraft were prohibited. In addition, the General Staff, seen by the Allies as one of the most important elements of the German military, was disbanded. This forced von Seeckt, the new German commander-in-chief, to undertake a thorough reorientation and evaluation of the organization. All the belligerents had gained experience in the First World War of the dilemmas posed by trench warfare and possible solutions to them. The Germans were the only ones who went a step further and subjected everything to a thorough analysis. In total, von Seeckt set up seventy-five commissions, headed by staff officers and consisting of experts in the various fields, to analyse the First World War in depth and breadth. More than 400 officers (10 per cent of the officer corps) were ultimately involved. This resulted in the aforementioned key document *Führung und Gefecht der Verbundenen Waffen*, one of the twentieth century's most important but least known texts in the field of military organization and doctrine. Part I appeared in 1921 and Part II in 1923. As its title indicates, a central theme was cooperation between the various branches of the army (infantry, cavalry, engineers, artillery, etc.) in integrated units. This document formed the theoretical framework for the further development of the Reichswehr in the interwar period. It would form the basis of the formidable military power that the world would witness in 1940.

The command concept described in *Führung und Gefecht der Verbundenen Waffen* fell on fertile soil. In the decades before the First World War, the Germans had formulated a number of important management principles for their military organization. Perhaps the most important was that of *Auftragstaktik*, formulated by von Moltke (1800–1891) in the middle of the nineteenth century. Evaluation of the German performance after the disastrous Battle of Jena in 1806 had shown that in battle there was little point in working with detailed plans: the smoke of the battlefield made central leadership impossible, and only local field commanders could properly assess combat situations on the spot. In the decades that followed, the Germans experimented with various command concepts until von Moltke introduced *Auftragstaktik* (*Führen durch Aufträge*) around 1860. The starting point of this concept was that at a central level planning of combat actions should focus on the main lines, and that local commanders then had the responsibility and authority to translate this into action at their discretion. In other words, the goals were clearly formulated centrally ('*what* should we have achieved *when*'), while the path to them was to be filled in at the local commander's own discretion within the framework of the current doctrine ('*how* are we going to achieve these goals'). This separation of the 'what' and 'how' resulted in a flexible organization with responsibility and authority exercised at a low level in the organization, and with a high degree of self-regulation. In other words, within *Auftragstaktik* the emphasis was on mission orientation and not on order orientation as in its opposite, centrally directed *Befehlstaktik*; German units were given a very high degree of independence instead of having to operate within the limitations of a detailed plan. They were also expected to have a high degree of flexibility, so that they could respond quickly to the opportunities and threats of the battlefield.

Auftragstaktik as a command concept did not fully come into its own in the First World War due to the extremely static nature of trench warfare, which did not allow for the flexibility that the Germans had in mind. One of the biggest problems in trench warfare was how to force and successfully follow up a breakthrough. If a breakthrough was achieved, the infantry and artillery, due to the difficult terrain, were usually unable to keep up with the rapidly advancing forward units and were easily isolated and neutralized by the enemy.

In the final phase of the First World War the Germans introduced *Hutiertaktik*, a variant of *Auftragstaktik*. In this, heavily armed units (*Stosstruppen*) infiltrated through the front line at key places (*Schwerpunkte*) and then quickly advanced into the enemy's rear to disrupt communications and destroy artillery positions. These infiltration techniques proved to be very successful. The problem remained, however, that regular infantry could not maintain the momentum of the attack: horse-drawn vehicles and supply columns could not keep up, giving the enemy the opportunity to regroup and seal the gap in the front. This weakness of the concept was to be overcome by the use of tanks and armoured tracked vehicles.

The new offensive doctrine of *Angriff im Stellungskrieg* (attack in static warfare) that emerged in this way fitted in well with the concept of *Auftragstaktik*: within a general framework, officers and men had to find their way without waiting for orders from above. This was in line with the fact that communications – in reality, telephone lines – were almost immediately shot to pieces by the artillery during battles, so that small groups of soldiers had to fight independently of each other and respond on their own initiative to opportunities that arose. In line with the new doctrine, an extensive training programme was set up to prepare men to use it. The German Army was in this way the only organization that recognized and acknowledged the limitations and challenges of the battlefields of the Western Front and responded to them by means of a new doctrine, training in leadership skills and delegation of decision-making to the lowest levels.

There is also a more cynical interpretation of this change in doctrine. In the hell of trench warfare after 1916, a different type of soldier had been born: a man with a closed face, imperturbable, unconcerned by the horrors around him, apathetic, but also creative, operating independently on the edge of insubordination. These men had learned to survive in small groups, without leadership and without discipline imposed from above. When no longer under the eye of his commander, only a man's own assessment of the situation could ensure his survival. This type of soldier needed a completely different form of command, if he needed one at all. The new tactical doctrine led to local successes in 1918, but could not avoid the strategic reality of growing Allied superiority in men and materiel, which ultimately led to the German surrender later that year.

More or less parallel to the development of *Führung und Gefecht der Verbundenen Waffen*, ideas developed about the possible applications of tanks. The First World War had shown that the most important problem after successfully forcing a breakthrough in the front was how to continue the attack with a second wave of regular infantry. Horse traction and the available vehicles in the First World War simply got stuck in ground made impassable by artillery bombardment and were unable to quickly follow up a breakthrough and consolidate it. Tanks and armoured personnel carriers could solve this problem. But as we have seen, between the first ideas about this subject and the application of them in Blitzkrieg, a lot of water still had to flow down the Rhine.

When finally, at the end of the 1930s, the new doctrine of armour slowly but surely took on more substance, an important difference with the First World War emerged: offensives were no longer focused on disabling the artillery but on destroying the C3I capacity (Command, Control, Communications and Intelligence) of the enemy. C3I can be seen as the brain and nervous system of the enemy, and if this could be damaged or destroyed, the enemy's front would collapse. Cooperation between the various army units in the mobile and hectic atmosphere of an attack and after a breakthrough was crucial to the success of this concept. The various branches of the army (tanks, infantry and artillery) had to be able to operate as integrated multifunctional combat units (*Verbundene Waffen*, or 'combined arms'). In other words, cooperation between these various army branches had to run smoothly at all times, without the traditional separation between them. Great emphasis was therefore placed on joint training and exercises, so that in combat there would always be optimal cooperation. Control of these multifunctional combat units was in the hands of one commander with ultimate responsibility, so that there was never confusion about who was giving orders. The commander had, in accordance with the principle of *Auftragstaktik*, as far as possible decentralized responsibilities and powers. By using a combination of tanks, mechanized artillery and infantry in armoured personnel carriers (the *Verbundene Waffen*), a new dimension was given to *Hutiertaktik* and a highly mobile method of warfare was introduced. As we have seen, Guderian, with his background as

The philosophical dimensions of combat: chaos theory versus planning and control

We must realize that this is a fundamentally different philosophy of battle. Almost all opponents that the Germans faced during the nineteenth and twentieth centuries saw it as their main challenge to bring order to the battlefield. They tried to get a grip on the complicated choreography of battle through centralized command and coordination, and attempted to fight along predetermined lines. In doing so, they were responding to the natural instinct to control one's environment and to organize it according to preconceived ideas. They therefore felt most at home on a linear and static front, which could be attacked at selected points with a calculated amount of men and resources. For the Allies in the First and Second World Wars, the ideal battle was almost a mathematical exercise, in which the correct tonnage of shells and an optimal mix of units at a given point on the front would physically pulverize the enemy. However, the more chaotic, fluid and obscure the battle was (or was made), the more the Allies lost their grip on events, sometimes leading to panic. To a greater or lesser extent, all Allied commanders had to deal with subordinates who asked, 'What should I do?' in unexpected situations – even in situations that actually offered them opportunities.

The Germans, however, accepted chaos as a natural aspect of warfare and were trained to deal with uncertainty. They used this confusion and uncertainty ('the fog of war') to gain advantage on the battlefield and tried to throw their opponents off balance by creating as much chaos as possible. By giving responsibility and authority to relatively low-ranking individuals, they enabled commanders to find their own way to pre-formulated goals within an agreed framework. A mobile, fluid, non-linear battlefield was the environment in which the German approach thrived. When this philosophy reached its maturity in the Second World War, in the form of Blitzkrieg, its key elements were: infiltrations, short intensive bombardments of communication centres, surprise attacks on the most unexpected places, isolation of the enemy's fortified positions, and the sudden appearance of mobile tank units in the enemy rear to destroy his logistical and communications structure.

a liaison officer, had also ensured that the units were equipped with sufficient radio and other communication equipment: an absolute prerequisite for maintaining cohesion between the units operating on a broad and deep front. The combination of *Auftragstaktik* and the deployment of multifunctional mechanized and armoured combat units formed the basis for the success of the German Army, especially in the early stages of the war. If all the above elements fell into place, this would result in Blitzkrieg, as displayed in the invasion of Poland in 1939 and the campaign of May 1940.

The tank offensive

The Germans realized that no opponent could be dominant over an entire front, but that, along the lines of the experiences of the First World War, forces had to be concentrated on one point, the so-called *Schwerpunkt*.[1] This was the weakest point in the enemy defence line, ideal in terms of terrain and attack routes, and offering good opportunities for artillery support.

The tank offensive concentrated on this part of the front. Its ultimate goals lay far behind the front line, and the tank units should thus advance as quickly and as deeply as possible. Under the concept of *Auftragstaktik*, commanders were given a great deal of leeway to act as they saw fit, provided that the set goals were achieved.

In the tank offensive itself, three phases were distinguished, namely:

- infiltration
- breakthrough
- pursuit

The infiltration phase

After a short preparatory artillery barrage and precision bombardment by the Luftwaffe, the tanks concentrated during the infiltration phase on the *Schwerpunkt* in the enemy's front line. The tank attack itself took place in two successive waves or two parallel groups. The task of the first wave was

---

1. This concept was actually formulated by A.H. de Jomini as early as 1803. All these innovative changes in the German Army were the result of the shock effect of the defeat by Napoleon in 1806 at Jena.

to break through enemy lines and eliminate the artillery positions behind them as quickly as possible, then advance deeper into the enemy's hinterland. Armoured infantry, the so-called *Panzergrenadiere*, always rode in the first wave in their armoured tracked vehicles, *Sonderkraftfahrzeuge* such as the SdKfz 251, to silence enemy guns. The tanks left the clearing of any remaining pockets of resistance to the second wave.

The firepower and speed of the initial attack usually broke enemy resistance, after which the units could quickly move on to reach their strategic objectives further behind the front. Through the breach thus created, infantry then poured as a second wave in trucks or on tanks, with the aim of keeping the gap open, protecting the flanks and lines of communication, clearing up pockets of resistance and capturing specific strategic positions. They were supported by anti-tank units to repel counter-attacks by enemy tanks. The ultimate goal was to give the third wave as much freedom as possible. The third wave's task was to eliminate the enemy's command structure in the form of his C3I capability. This third wave itself was to remain intact as much as possible, because the greatest damage was inflicted on the enemy in this phase.

The breakthrough phase

If the enemy's command structure in the form of C3I capacity was successfully eliminated, his organization would collapse and coordinated resistance would no longer be possible. In the vacuum thus created, the German armoured troops, with their ample C3I capacity, were able to advance deep into the enemy's rear, leaving the mopping-up of pockets of resistance to the infantry units that had been advancing behind the first waves of attack.

The pursuit phase

This began at the moment that the enemy began to retreat or flee en masse. The aim of this phase was to prevent enemy troops from regrouping by continuing to chase them, using rapid manoeuvres, supported by the Luftwaffe, to penetrate further into the areas behind the front. The ultimate goal was to neutralize all resistance.

Compared to a traditional engagement, in which the destruction of the enemy's troops was the main focus (*Vernichtungsgedanke*), destroying his C3I capacity was an effective and rapid way to bring an enemy to his knees at relatively low human cost (for both the attackers and the defenders). Bearing in mind the horrific slaughter of the First World War, the aim was not to destroy men and equipment, but to deprive the enemy of the opportunity to use them. Speed, mobility (*Beweglichkeit*) and the creation of chaos were paramount in the tank offensive: this gave free rein to the superior training and leadership of German armoured troops.

An important innovation, compared to the First World War, was the use of the armoured half-tracks of the *Sonderkraftfahrzeuge* family, the SdKfz 250 and 251 and their variants, alongside the tanks. These vehicles enabled the Panzergrenadiere to keep up with the pace of the tanks and avoided the main reason for the previous failure of tank offensives, namely the separation of tanks and infantry (with their anti-tank guns) at the moment of attack. If for some reason the half-tracks could not keep up with the tanks they remained under cover until the moment when the attack could be resumed together. The half-tracks not only protected the soldiers from enemy fire, but were also perfectly capable of operating off-road, an essential ability in areas where good infrastructure was lacking.

Reconnaissance in force

An important aspect of these new tactics was so-called 'reconnaissance in force'. The Germans attached a much higher value than their opponents to gathering information through reconnaissance, and from the beginning, separate reconnaissance units were set up and trained for this purpose.

The Germans believed in fighting for information, so the reconnaissance units were not only highly mobile and well-equipped, they were also very heavily armed and armoured. Each division had such a reconnaissance battalion, the *Aufklärungsabteilung*, whose task was not only to gather information, but also to take advantage of any opportunities that presented themselves during armed reconnaissance actions. This *Aufklärungsabteilung*, often the most mechanized and heavily armed unit of the division, also served as a strategic reserve for the less motorized divisions. In addition to general information about gun positions, etc., reconnaissance had to provide information about any weak spot in the front which could serve as the *Schwerpunkt* for an attack.

During reconnaissance actions, the units scoured the enemy front in a fluid operation in search of these weak points. Due to the great firepower of the reconnaissance units and the intensity of the skirmishes they provoked, these actions were often seen by the enemy as an actual attack. Observation of the disturbance this caused on the enemy side, in the form of counter-fire, troop movements, radio traffic, etc., provided useful information for possible follow-up actions. Once a *Schwerpunkt* had been identified, the reconnaissance unit would contact the divisional commander. The latter could then either initiate the attack by sending in armoured units or order the reconnaissance unit to continue its own attack and force a breakthrough. In this way, reconnaissance actions on a section of the front could suddenly turn into a regular attack with much greater tactical implications. Reconnaissance units were therefore a factor that the enemy had to take into account at all times.

All in all, based on experiences in the First World War, the Germans formulated a clear vision of their military organization and a distinctive method of operating. The German Army at the end of the 1930s could to a certain extent be seen as a project organization. When the units were not in combat, a panzer division, for example, was structured along the functional lines of a tank regiment, artillery unit, infantry unit or supply unit. If they went into battle, Kampfgruppen – in essence, large project teams – were deployed and given an assignment (*Auftrag*) with certain objectives. These Kampfgruppen were very diverse in composition and size and functioned until their objective was achieved, after which they broke up into the various functional units or were deployed as a new Kampfgruppe. In this way, a particularly fluid organizational structure was created, built around Kampfgruppen, that could vary from day to day: a project organization in the true sense of the word. It is clear that this aspect of *Verbundende Waffen* placed very high demands on both organization and personnel, but because the Germans were constantly working to anchor this concept in the breadth and depth of their organization, they were able to make it work in practice.

The German Panzer Division

The panzer division was a central element in this concept. The standard structure of a German panzer division in 1940 is shown in the diagram below.

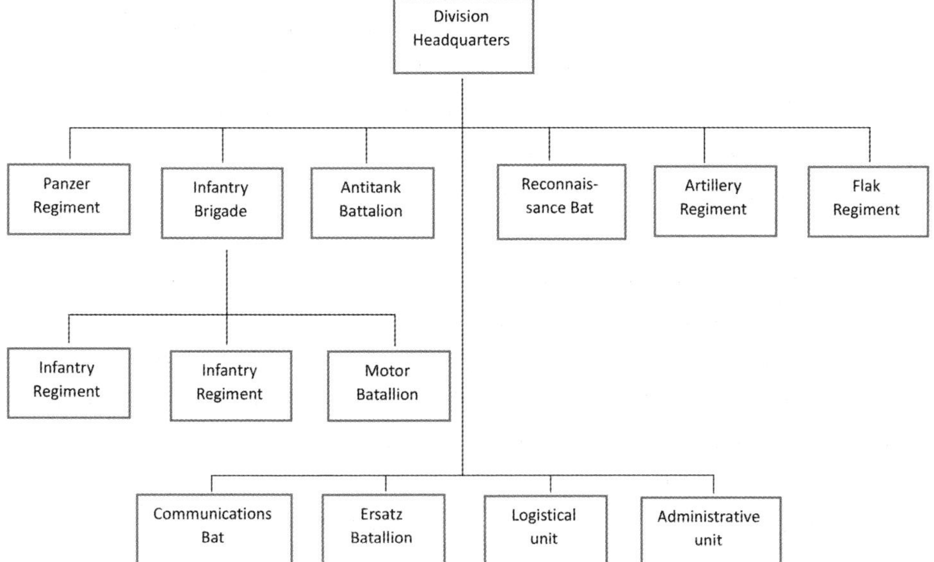

Structure of a German Panzer Division 1940: 1st Panzer Division.

The core of the panzer division was formed by the panzer regiment, structured as follows:

- Regimental Staff, with
  - o *Nachrichtenzug* (communication unit)
  - o *Leichter Panzerzug* (light tank platoon, for carrying out reconnaissance)
- Two Panzer *Abteilungen* (tank battalions)
- Panzer *Werkstatt Kompagnie* (field workshops company and salvage units)

The two Panzer *Abteilungen* were constructed as follows:

- *Abteilung* Staff with
  - o *Nachrichtenzug*
  - o *Aufklärungsabteilung* (reconnaissance unit)
- *Leichter Panzerzug*

- *Pionierzug* (engineer platoon)
- *Fliegerabwehrzug* (anti-aircraft platoon)
- Two light companies (equipped with the PzKpfw III)
- One medium company (equipped with the PzKpfw IV)
- One light (supply) column

The tank regiment could attack with both its *Abteilungen* one after the other in two waves, or side by side. The *Abteilungen* were made up of three companies of between twenty and twenty-five tanks each, depending on how many were operational at a given time. They were deployed in triangles, the so-called *Keil* (wedge) with the point of the triangle forward, or a *Breitkeil* (wide wedge) with the base of the triangle forward. The companies within the triangle were positioned during the attack in such a way that all tanks had a maximum field of fire and could cover each other as much as possible. When one tank failed, another took its place, so that the formation was maintained for as long as possible.

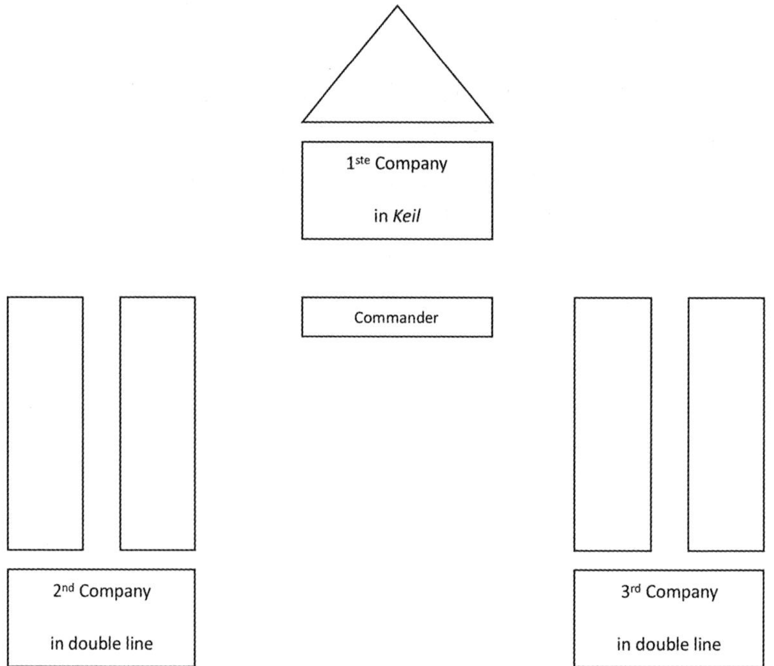

Attack formation of a German Panzer Abteilung in so-called *Keil* (wedge).

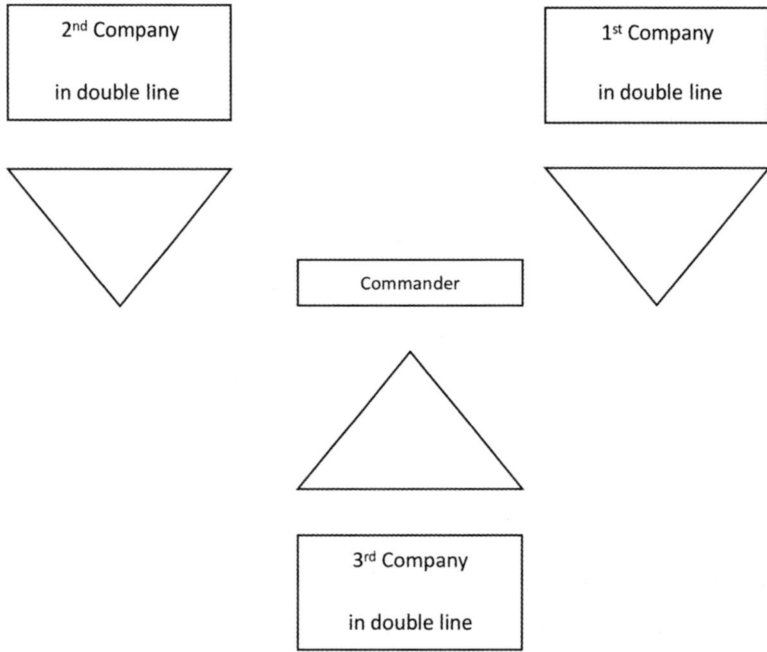

Attack formation of a German Panzer Abteilung in so-called *Breitkeil*.

In accordance with the principle of *Verbundene Waffen*, an infantry brigade with two regiments and a motorcycle battalion, an artillery regiment and an anti-tank battalion formed an integral part of the panzer division. A division contained approximately 10,000 men.

These divisions would form the core of the German Army in the Second World War and initially bring them great success.

Chapter 6

Developments in Other Countries in the Interwar Period

Germany was not the only country which grasped, after the First World War, the possibilities that motorization in general and the tank in particular would offer to the future waging of war. Similar discussions took place in Great Britain, the Soviet Union, the United States and France, and we will examine these countries in turn

Great Britain

Tactical doctrine

To get an idea of British tactical doctrine and the role of the tank within it in the Second World War, we will first look at the period before and during the First World War. British tactics before the First World War were mainly based on experience in relatively small-scale wars across the British Empire, so operating in larger military units was neglected. Morale was seen as an important ingredient of success; in *Infantry Training* (1914) the aim was stated as being 'to close with the enemy, cost what it may'. W.R. Robertson stated in January 1914 that commanders should never 'interfere with the desire of the troops to push into fight at all costs'. The underlying idea was that any future war would be short and that decisive victory would be achieved quickly. As the First World War dragged on and the size of the British Army on the Continent increased, the high command increasingly fell back on solutions from the colonial past. As a result, the dominant British tactic of the First World War consisted of an initial artillery bombardment followed by a massive infantry attack with the bayonet. The attack had to be continued at all costs. The key to success, in the eyes of the British high command, was the deployment of ever-increasing numbers of men and equipment, until the enemy finally succumbed. Needless to say, this cost an unnecessarily huge number of lives. The losses speak volumes: compared to 1.8 million dead on the German side, no fewer than 3.8 million French and British were killed.

After the First World War, neither the British nor the French, as victors, felt the need for a fundamental analysis of the war's experiences and dilemmas. Existing tactics were not further elaborated, and no new ones were developed to meet any problems that were identified. It was not until fifteen years after the war that a report, *Notes on Certain Lessons of the Great War*, was drawn up. The report concluded that surprise was an important element in success and that the British had acted too stereotypically in all kinds of situations. In the future, commanders had to take more initiative themselves, and questions were raised about the usefulness of prolonged preliminary artillery bombardments: they gave away the element of surprise and destroyed the ground to such an extent that it was difficult if not impossible to carry out an attack successfully.

Regarding the use of tanks in a future conflict, a lively discussion took place between the two world wars. The publications of people like Fuller and Liddell Hart stimulated thinking about the subject, but this did not lead to a generally accepted, broadly supported doctrine.

Development of tank theory in the interwar period

After the First World War, F.C. Fuller was the first to develop the idea of future warfare in which the tank occupied a central role. As a staff officer, Fuller had drawn up the 'Plan 1919' in 1918. Containing the core of his later ideas in the field of mobile warfare, this was based on a frontal attack with 1,592 (!) heavy and 390 medium-weight tanks charged with breaking through the front at various locations. The breakthrough was to be followed by a second wave of attacks by 790 medium-weight tanks aimed at destroying the enemy's command centres, and 1,220 medium-weight tanks, which were to clear up the last pockets of resistance. The plan never got further than the paper it was written on, as neither the types nor the numbers of tanks were available. However, Fuller developed the concept described above further in the 1920s, based on the following ideas:

- The introduction of the internal combustion engine will lead to a revolution in the field of warfare
- The tank is the basic building block of every army
- The tactical doctrine of a rapid breakthrough and a subsequent encirclement should replace the existing tactic of a frontal attack by infantry

Fuller's ideas were supplemented by those of a young infantry officer, Captain B.H. Liddell Hart. Based on his experiences in the First World War, he advocated concentrating forces on a limited part of the front with the aim of forcing a breakthrough there. Once the breakthrough had been achieved, the attacking units should advance further, but without losing contact with those following behind them. This avoided two problems: on the one hand, that following units which were too slow could easily be cut off from the attackers; while on the other hand, that waiting too long to deploy following units would lead to the attack losing momentum.

When combined, these ideas resulted in the late 1920s in a new concept: the tank as the central element instead of the infantry. It was based on the speed and firepower of tanks, combined with infantry in armoured personnel

The Matilda II tank was designed to support infantry. It was slow, weakly armed but heavily armoured; only the Flak 88 gun could stop it. Rommel's units first destroyed Matilda tanks with this type of gun at Arras in 1940; in North Africa, history would then repeat itself.

carriers – a concept, of course, that had many similarities with ideas current in the Reichswehr. Experiments were carried out, but after some initial success a follow-up was decided against, and a deadlock arose in discussions about the deployment of armoured units. There was no consensus in the high command about the future role of tanks, and the proponents of armoured units were unable to jointly develop concepts about the way in which these units should be organized and deployed.

However, the mechanized 7th Infantry Brigade and the 1st Tank Brigade, the two most modern units of the British Army, were combined in the early 1930s to form the so-called Mobile Force. Disappointingly, this formation proved to be no match for a conventional infantry division during exercises in 1933. The reasons for this were poor planning, execution and coordination, problems that would continue to haunt the British during the Second World War. This failure did nothing to help the cause of mobile warfare, and after lengthy discussions, it was finally decided in 1935 to equip one cavalry division with tanks and to rename it the Mobile Division. In the years that followed, successive commanders of this division developed neither a vision nor a doctrine concerning the role of the tank. They were torn between two ideas: either tanks as a 'cavalry' screen in front of the line, or as a unit charged with forcing a breakthrough.

When the British Army started mechanizing on a large scale in 1938 they did not have a clear idea of the purpose for and manner in which they wanted to deploy the newly formed tank units. The result was that Britain entered the war in 1939 without a coherent vision of mechanized warfare and without a single effective armoured division. Nor had any thought been given to coordination between infantry and tanks, let alone coordination between the army and the RAF in the form of ground support by the latter. There was also an impasse regarding types of tank until the mid-1930s. Ellis, as Master of Ordnance responsible for the procurement of weapons systems, could not agree with the vision of Fuller and Liddell Hart about the shape of a possible future war and the role of the tank in it. He saw the tank only as supporting the infantry, and a role for light tanks in the field of reconnaissance. He was not interested in heavier tanks with sufficient speed and firepower to be deployed as a relatively independent fighting force. In line with this, all attention was focused on the production of infantry support tanks, the ultimate result being the Infantry Tank Mark II, or Matilda, from Vickers.

This tank was heavily armoured, equipped with a 2-pounder (20mm) gun and had a crew of four. The Matilda had a top speed of only 22.5kph, as it had to operate together with the infantry. Vickers also supplied the Vickers Medium tank as a light reconnaissance vehicle. In addition, impressed by the performance of the Soviet BT 26 tank, several types of so-called 'Cruiser' tanks were developed in the mid-1930s. This type was faster than the infantry tank and better armoured than the reconnaissance tanks. The end result was that the British Army in 1939 had three types of tanks at its disposal, namely:

- light tanks for reconnaissance purposes
- infantry tanks for supporting the infantry
- cruiser tanks for use in tank formations in a more independent role

Contrary to the original ideas of Fuller and Liddell Hart, and those of Guderian, British tanks operated either as fast, lightly armoured independent units without support from motorized and armoured infantry (the 'Cruiser' family), or as slow, heavily armoured infantry support units (the 'Matilda' family). These tanks were organized into two Mobile Divisions, one for use on the Continent and in Britain itself, and one in Egypt. The further development of the doctrine of mechanized and mobile warfare was left to these divisions with their cavalry background. They proved unable to explore the potential of the tank, let alone to develop it through training and battle simulations. The end result was that at the outbreak of the Second World War no common doctrine had been developed on the use of the tank in a future conflict.

The blame for the lack of a clear tank doctrine must certainly not be placed solely on the cavalry regiments; it was also the fault of men at the top of the British Army. General Paget, Commander in Chief Home Forces (CIGS), declared: 'Anyone can handle armoured forces. No special knowledge is needed.' Regarding cooperation between armoured and infantry divisions, it was noted as late as in November 1942 that 'operations between an armoured division and one or more infantry divisions had not been studied and had certainly not been practised' – despite the fact that in preceding

The place of the tank

In the 1920s, the British Army's leadership attempted to reduce the number of regiments. For this reason, the new tank weapon was assigned to existing cavalry units, rather than to the Royal Tank Corps. During the 1920s and 1930s, the cavalry regiments battled the high command over the question of exchanging their horses for tanks. When the cavalry regiments were finally equipped with tanks they continued to view tank tactics from the perspective of a cavalryman, completely ignoring the fact that the tank was a completely new weapon system. In line with tradition, these cavalry regiments were assigned the Cruiser tanks in order to be able to operate as an independent fighting force. Units of the Royal Tank Corps were attached to infantry units and equipped with Matildas. An example of the attitude of the cavalry regiments comes from one of the first encounters between British tank units and German armour in North Africa, when an officer 'had seen the Hussars charging into Jerry tanks, sitting on top of their turrets more or less with their whips out.' The commentator remarked dryly that 'they had incredible dash and enthusiasm, and sheer exciting courage which was only curbed by the rapidly decreasing stock of dashing officers and tanks.'

It is striking that apparently the British had learned nothing from the experiences of May 1940; after all, it was not the first time that these parties had met.

years it must have been clear that the success of the German Army could be explained in no small part by the joint deployment of infantry and tank units. Such lessons seemed to have completely bypassed the leadership of the British Army. Lord Carver, a junior officer during the Second World War and later commander-in-chief of the British Army, recognized the weaknesses in British doctrine:

> Our real weakness is our failure to develop a doctrine for a concentrated attack with tanks, artillery and infantry on a limited part of the front. Again and again our tanks attacked on a broad front, the momentum of the attack being immediately lost the moment the front line was brought to a halt by enemy tanks or anti-tank guns. If artillery was involved in the attack, it then

occupied itself with bombarding enemy positions, after which the next line of tanks moved into line again to suffer the same fate. Infantry played no part; it served to occupy objects after they had been captured by the tanks.

This was, in a nutshell, the British way of operating. These tactics had not changed significantly at the end of the Second World War. At its outbreak, there was no clear concept in which tanks played a role in breaking through the enemy's lines. As a result, the British method of operating most closely resembled the tactics of the First World War: after an artillery bombardment, they attacked on a broad front in the hope that the guns had sufficiently undermined the enemy positions.

During the war, the British would also learn little from their own experiences and those of the other belligerents. As a result, the use of tanks and the integration of the various army units and weapon systems was and remained one of the greatest operational problems of the British Army during the Second World War.

The Soviet Union

In Search of a Vision[2]

The Soviet Union had other concerns in the 1920s. It was recovering from a bloody civil war, the painful defeat by Poland and the resulting Soviet-Polish Peace Treaty of Riga of March 1921, which had awarded to Poland large areas inhabited by White Russians and Ukrainians. This had put an early end to Soviet ambitions to convert Poland to international communism. The new border had cut the Soviet Union off from Germany and Lithuania and would remain in place until 1939. The 'Red Workers and Peasants Army', the official name of the Soviet Army, was ready to reflect on its role. According to leading Marxist political/strategic theorists, it was clear that the internal contradictions of capitalism meant that the coming of war was only a matter of time. Coalitions of states would be involved, and the war would most likely become global. It was realized that technological developments would influence the battlefield of the future, and strategies were developed to make

2. This section is based on the publications of M. Garder, G.M. Glantz, S.J. Main, O'Balance, L. Samuelson, H.F. Scott, A. Seaton, S.W. Stoecker, and E.F. Ziemke.

the most of these technologies. This sense of the threat of war developed into a real psychosis in 1927: Great Britain had broken off diplomatic relations with the Soviet Union, the Soviet ambassador in Poland had been assassinated and there had been attacks on Soviet representatives in Berlin, Peking and Shanghai. Stalin declared that war was inevitable, although it was still unclear against whom it would be fought. Based on research that had begun under Tukhachevski in 1926, it had emerged that the Red Army needed:

- motorized infantry machine gun units, supported by large formations of fast tanks and motorized artillery
- large cavalry units, which had to be reinforced with armoured units (armoured cars and fast tanks) and could have great firepower (optimal utilization of machine guns)
- large units of attack aircraft

A report detailing these requirements and published in the midst of the war psychosis had a sobering effect on the political leadership, since the Red Army had none of the above-mentioned weapon systems. One of the consequences of the war psychosis of 1927 and Tukhachevski's study was that Stalin became convinced that preparations for a future conflict had to be accelerated. Under the leadership of Commissar for Defence Voroshilov, a Five-Year Plan was drawn up in 1928 which was to increase the number of tanks from 380 to 7,000, aircraft from 1,232 to 3,332 and guns from 999 to 4,870 by 1933. Official figures show that in 1933 the intended objectives had been largely achieved: the number of aircraft was at the desired level, more than 10,000 pieces of artillery were available, but the production of tanks, with only 1,053 units, was clearly behind schedule. The question now was how to deploy them.

Soviet military doctrine: the theory of deep penetration

In the early 1930s, many people in the military hierarchy were thinking about the shape of a future war, the role of new technologies in it and the way in which these could be most quickly incorporated into the Red Army. Tukhachevski had become one of the leading army officers and director of the Frunze military academy. His ideas dominated all other initiatives, and in 1931 he became Deputy Commissar for naval and military affairs. In the autumn of 1931, Tukhachevski unveiled his plans to the Politburo for the reorganization of the Red Army, plans that had gained urgency due to political developments in Germany. The so-called 'Tukhachevski Plan' was to lead to a number of radical changes in the Red Army over the next three years. These changes not only concerned its organizational structure, but also included the absorption of the territorials into the regular army and the creation of motorized and mechanized units. The latter changes were in line with the theory, generally accepted in the countries of Western Europe, that victory in a future war would be determined by the success of the first attack, which would have to take units deep into the enemy's rear. There they would have to destroy his entire logistical and C3I system, resulting in the collapse of enemy resistance. The Soviet doctrine of deep penetration was based on breaking through the enemy's defence line, followed by an advance of 150–200km deep into the enemy's rear. The Soviet units would have to advance in a wide wedge at a rate of between 10 and 32 kilometres per day. Tanks, aircraft and shock armies were central to this concept and these had to advance faster than the 'bourgeois' units could retreat. Tukhachevski assumed that three types of tank were needed for this tactic:

- to support the infantry in capturing fortifications
- to support the infantry in the open field
- to operate as an independent tank force deep behind enemy lines

In other words, Soviet views were similar to those of the Germans on a number of important points. The first type of tank was to assist in breaking through the defensive line; the second type was to support the infantry in exploiting the gap created; and the third could then pour through the gap at high speed towards targets located far in the enemy's rear. Meanwhile, the air force would bomb the enemy's supply lines and communication centres, and paratroopers could be deployed against specific targets. By 1936, the Soviets had established four mechanized corps on the basis of this doctrine, each consisting of 500 tanks and 200 armoured vehicles. In addition, a special organization was built up around ground support by the air force of these deep penetrations behind the enemy front line. Although the doctrine was still under discussion, the new officer manual of 1936, *PU-36*, was entirely built around this doctrine of deep penetration. An 'Instruction for Deep Penetrations' had also been drawn up for the army top brass. Themes within *PU-36* were rapid manoeuvring, close coordination between the various weapon systems (*Verbundene Waffen* in German terms) and bringing the war into enemy territory. According to this manual, 'the infantry, in cooperation with artillery and tanks, will determine the outcome of the battle. For this reason, the other units carry out their mission in support of the infantry.' There was also room for cavalry: 'strategic cavalry [is] capable of operating independently under all circumstances.' Tanks could operate independently, but one had to take into account 'the technical limitations of the material'. With this document, Soviet ideas connected with the generally accepted doctrine formulated by the Reichswehr in 1933 and published in 1936.

The Soviet Union had also produced a series of tanks with a longer range and higher speed in the early 1930s, the BT (*Bystrokhodny* [high-speed] Tank),which was initially armed with a 37mm gun, later with a 45mm and 76mm version. There was also a lot of experimentation with other types of tanks, including amphibious ones, and armoured personnel carriers. The Red Army also grew in size. In 1931 there were thirty regular and forty-one territorial divisions, but this number began to grow rapidly, and by 1934 the army had increased in size from 600,000 to 940,000 men. There were two types of armoured unit: the tank brigade and the mechanized brigade. Tank brigades consisted of four battalions, each with thirty heavy tanks, mostly T-28s of 29 tons, which had been mass-produced during the second Five-Year Plan. Mechanized brigades were larger and more flexible, with five battalions, three equipped with light tanks, one with reconnaissance tanks and one with machine guns on armoured vehicles. The T-26 and the BT-5 became the standard tanks for these brigades.

Soviet artillery showed a more diverse picture. There had been experiments with numerous variants for anti-tank, anti-aircraft and artillery guns. After some hesitation, the Red Army went for a basic artillery battery of six to eight field guns; there were 1,500 of these batteries. There were shortages of ammunition, however, because no decision had yet been made in the field of standardization. The majority of the artillery used horse traction, but here too mechanization was increasing: the production of tractors for

A T-26, Model 1931, drives through a river near Moscow in 1936. One of the crew members anxiously watches to make sure the hull is indeed watertight.

military use was part of the second Five-Year Plan, and in 1936 the Red Army already had 150,000 (!).

The infantry divisions were large, consisting of about 19,000 men and made up of three regiments, which each consisted of three battalions. There were also divisional units of artillery and engineers, as well as reconnaissance and support units. These divisions also became increasingly mechanized, with trucks being used to transport troops and horse and cart to transport supplies.

The Soviet air force had initially copied American and British designs, and in 1935 they had between 4,000 and 5,000 aircraft at their disposal. In addition, they produced their own impressive four-engined bombers, slow machines but with a huge load capacity. Moreover, the designers and pilots gained a lot of experience due to the great distances and the different and extreme weather conditions of the Soviet Union. Once a certain type of aircraft was introduced, massive numbers of pilots were trained, enough to build up a substantial reserve. This culminated in an air show during the

May Day Parade of 1935 with no fewer than 3,000 aircraft. The air force was prevented from acting too independently: it served mainly to support ground troops in achieving the army's strategic goals.

In 1936, during large-scale manoeuvres, the doctrine of deep penetration and the principles of *PU-36* were tested in practice. Six infantry and four cavalry divisions, together with 1,200 tanks and 1,000 aircraft, participated in these exercises. British and French international observers were deeply impressed by the size of the force, but less so by its tactical capabilities. The general opinion was that the Red Army would be a formidable opponent on its own territory, but would not be able to wage a war of aggression against a European opponent. This was, incidentally, only one of the more than 140 exercises that were held in the period 1936/1937.

Tukhachevski, however, signed his death warrant with his views on the position of political commissars within the armed forces, suggesting that their role had to be reduced as much as possible given their limited military expertise. This and other proposals by Tukhachevski ran completely contrary to the party line and very much contradicted the wishes of Stalin and the hardliners. When Tukhachevski published an article about *PU-36* in the army newspaper *Red Star* and the party newspaper *Bolshevik* in May 1937, this was used as an opportunity to remove him. On 10 May, a decree was issued that fully reinstated the powers of the political commissars. Tukhachevski was relieved of his post and arrested by the NKVD on 26 May, then executed on 11 June. With this, the Red Army was deprived of its most important

This T-35A (mid-production) from 1937 shows the Soviet interest during this period in true 'land battleships': no fewer than four gun turrets aimed to give this mastodon massive firepower. Fortunately for the Red Army, there were also more practical designs such as the T-34 and the KV-1, which would play a decisive role in the future conflict with Germany.

visionary leader. The subsequent purges in the Red Army, directed primarily against former Tsarist officers and civil war veterans, had a devastating effect.

The removal of Tukhachevski and the group of officers around him marked the end of the creative and progressive school of thought in the Red Army. As tanks, guns and aircraft continued to pour off the assembly lines, they were distributed without vision among the various army units. Initiative

The Spanish Civil War

The Spanish Civil War offered the Soviets a good opportunity to test their weapon systems in war. The Soviets provided the Republicans with tanks, planes and artillery. A limited number of people went with them; there were never more than 500 to 600 Russians in Spain. Most were staff officers, political and intelligence officers or instructors accompanying the tanks and planes. Their mission was to see how the Soviet tanks, artillery and planes performed in combat, as well as to organize the Republican forces. They were instructed not to become involved in combat or in situations where they could be taken prisoner. Of all the officers who visited Spain, Pavlov, the tank expert, was the most important. He brought the first shipment of fifty tanks to Spain in October 1936, stayed there until 1938 and led several Blitzkrieg-style tank actions. For example, in the Battle of Guadalajara, the first tank battle of modern times, he led seventy Soviet tanks against two divisions of Italian tanks. The Italian tanks performed poorly and were blown apart by 150 Soviet aircraft. German observers watched from a distance, saw the Italians' weaknesses and drew different conclusions to Pavlov: he saw no role for large, independently operating tank units; in his eyes they should be combined with the infantry. This was the same view as that of the French and to a lesser extent of the British. Back from Spain, Pavlov's experiences were useful in discrediting Tukhachevski, with the consequences described above. Soviet aircraft designs were also evaluated. The Soviets found them to be excellent compared to German and Italian aircraft. And not unimportantly, they defined the support of ground troops and not strategic bombing as the main future mission of the air force.

was discouraged, and only standard approaches and solutions were accepted. Officers were nervous about being the next to be purged, and became uncertain of their role and position now that professional competence was no longer as important as ideological conformity.

Tukhachevski's ideas were replaced by the views of D.G. Pavlov, a tank expert with experience in the Spanish Civil War. In his eyes, tanks had had a great morale-boosting effect on the infantry, but had not been successful in independent units. In his view, tanks were there only to support the foot soldiers. This view became the Red Army's guiding principle, the motorized and mechanized units were disbanded and the tanks were dispersed among the various divisions.

In fact, with the abandonment of the doctrine of deep penetration, a vacuum appeared, which became obvious when the Red Army struggled to overcome Finnish resistance in the Winter War of 1939/1940.

Meanwhile, the second Five-Year Plan had borne fruit. More than 9,000 tanks became available, mainly T-26s and the BT series. Up to now, the Soviets had copied tanks of British and American make, but now for the first time they were able to develop their own: the T-28, the M-2, the T-34 and the KV-1. In addition, some 2,000 amphibious vehicles, armoured or not, were operational.

The number of artillery pieces had grown according to plan to 7,000, the majority of which were pulled by armoured tractors. Anti-tank guns, anti-

One of the most unpleasant surprises faced by the Germans would be the KV-1. This heavily armoured and armed tank was almost impossible to knock out, even with the Flak 88 gun, and engineers often had to use explosive charges. The text on the gun turret reads: 'Victory is ours'.

aircraft guns and mortars were also either pulled by lorries or transported on them. Some thought was already being given to mechanized artillery guns that were fully armoured. Ideally, this artillery, combined with the tank and mechanized brigades, would form a single entity that would move as a single armoured unit across the battlefield of the future.

All in all, by the time war broke out, the Soviet Union had acquired an impressive military force that far exceeded that of countries such as France, Great Britain and the United States. Only Germany could match it to some extent. In 1939, the moment of truth would arrive for the Soviets: the Polish campaign and the Winter War against Finland would reveal both the strengths and the weaknesses of the Red Army.

The United States

In line with the reduction of the American Army after the First World War, the General Staff disbanded the Tank Corps, and under the National Defence Act of 1920, tanks and tank development, the latter in consultation with the Ordnance Department, now came under the infantry.

The General Staff described the role of the tank in a future war as 'to facilitate the uninterrupted advance of the rifleman in the attack'. The Americans had 100 British Mark VIII tanks at their disposal which were mainly assigned to the 67th Infantry (Tank) Regiment. They were used for exercises and deployed as support for the infantry. After the introduction of new American tanks in 1932, they were gradually taken out of service. The Americans also used the light Renault FT tank. Under the name M 1917, 950 of these 6-ton vehicles would remain in service with the Americans until the 1930s. Eventually, some of them ended up with the Canadian Army in 1940.

For economic and operational purposes, tanks were divided into 'light' and 'medium' types; the former had a maximum weight of 5 tons and had to be transportable by truck, the latter were not allowed to exceed 15 tons, given the weight that military bridges could carry. Walter Christie regularly pestered the Ordnance Department to adopt his own light and fast tank designs, but apart from the purchase of a few of his tanks, his views were not listened to; the Ordnance Department followed its own ideas in the field of tanks. Serious financial constraints in the 1920s limited tank development to about two experimental models per year, culminating in 1931 in the development of the T1E4 light tank. This had the engine in the rear compartment and the drive shaft in the front, and this design would become the basis for all American light tanks.

In 1927, the US General Staff set up a small Experimental Mechanized Force, following the example of a similar unit in Britain. This Experimental Mechanized Force was made up of the tank units of infantry divisions and used light tanks. In 1931, General Douglas MacArthur decided that in the age of mechanization, the US Cavalry, equipped with tanks and armoured cars, should fulfil the role of exploiting possibilities on the battlefield. His vision of the future place of the tank differed significantly from its traditional role in the US Army up to that time. The cavalry took over the Experimental Mechanized Force and was allowed to equip itself with tanks. In order to stay within the National Defence Act, the cavalry tanks were called 'combat cars', a legal fiction that avoided the rule being broken that only the infantry could use tanks. A similar scheme will also be encountered in France.

By 1934/1935, several experimental tanks had been produced: the T2, the T2E1 and the T2E2. The T2 itself was inspired by the British Vickers Armstrong 6-ton tank, including the leaf springs of the rear suspension. For economic reasons, it was desirable that the cavalry 'combat cars' should be derived from the light infantry tanks, and in line with this, Rock Island Arsenal produced a similar tank for the cavalry, the T5 combat car. After the necessary modifications, this tank entered service with the US Cavalry in 1937 as the M1 Combat Car. In July 1940, due to developments in Europe and the victorious march of the German armoured forces, the American Armoured Force was established, making the distinction between cavalry and infantry tank units a thing of the past. The term 'combat car' was no longer necessary, and the M1 and M2 Combat Car were renamed Light Tank M1A1 and Light Tank M1A2. These tanks would never see combat, but would primarily be used in training. However, the M1 and M2 Combat Cars would form an important basis for future American tank designs until 1944, and the US Army was able to gain much development and operational experience by using them.

American views on the role of tanks remained limited to the support of infantry. They did not aim, as the Germans and Soviets did, to use tanks

An American M2A3 light tank at speed during war games in 1940. This type of tank would not ultimately be deployed at the front.

for deep penetration into the enemy's rear. In 1940, the United States simply did not have tanks that were suitable for the task, and even later in the war, the Sherman, their heaviest tank, would not change this. The Americans were betting on achieving superiority in men and materiel, and their use of tanks would remain similar to that of the First World War. How vulnerable American tanks were compared to their German counterparts is shown by the fact that the ratio of Panthers to Shermans lost was 1:10: for the destruction of each Panther tank, ten Shermans were knocked out, and consequently, a breakthrough by American tanks could quickly be halted by German tanks and anti-tank guns.

France

A battle of ideologies

Developments in France do not generally come under the spotlight as much as those in other countries, but similar discussions about the tank's role were held there, and various interesting tanks and armoured vehicles were created that are worth a closer look. The French made a number of fundamentally different choices to the Germans in the construction and deployment of their tank forces, the origins of which lay in the First World War. The deployment of tanks in the war had made a great impression on the Germans. In the opinion of many German analysts, the surprising deployment of large numbers of tanks at places on the front where they were not expected had made a major contribution to the Allies' victory. The French, on the other hand, were much less convinced of this and, tired of the war, devoted little energy to the further development of ideas about the deployment of the tank. As the most mechanized army of the First World War, they saw no other roles for the tank than those of supporting the infantry and acting as reconnaissance units to replace the cavalry. They assumed that offensive actions would develop along the same lines as in the First World War and entrenched themselves behind the fortifications of the Maginot Line. They could not imagine tanks playing a role in mobile warfare and thus envisaged them mainly operating in small units (between thirty-three and forty-five tanks) with infantry divisions.

But France, like Germany, Great Britain and the Soviet Union, also had proponents of the tank as an autonomous weapon. The most prominent of these were Jean-Baptiste Eugène Estienne and Charles de Gaulle. After the war, Estienne was until 1927 commander of the French tank forces, first as commander of the Artillerie Spéciale, and later, when all tanks were integrated in the infantry, Commander of the Chars in 1920. In 1919, Estienne presented Pétain with a proposal entitled *Study on the Missions of the Tank*. This document emphasized the importance of armoured tracked vehicles for the transport of infantry, artillery and recovery teams that operated in combination with tanks and were supported by aircraft that would carry out bombardment deep in enemy territory. This concept was far ahead of its time and was comparable to the ideas of Tukhachevski in the 1930s. At a conference in Brussels in 1921, Estienne advocated a force of 100,000 men, equipped with 4,000 tanks and 8,000 transport vehicles, which could break through an enemy's front and advance 80km in a single night. However, these ideas fell on deaf ears, and the French military establishment in the 1920s focused entirely on the infantry as the key weapon in a future war.

Although politicians such as Paul Reynaud supported Estienne's progressive ideas and advocated a mobile force as early as 1924, they were in a minority. French military doctrine saw the role of the tank only as supporting the infantry, something that would not change until Weygand became commander-in-chief. Within the army, Charles de Gaulle, an enthusiastic supporter of Estienne, was particularly interested in the potential of the tank to support rapid mobile actions, with the ultimate goal of avoiding the slaughter of trench warfare. In the 1930s he wrote several books and articles on military subjects that showed him to be a sharp analyst and a gifted writer. In 1931, he published *Le fil de l'épée* (*The Cutting Edge*), an analysis of military and political leadership. He also published *Vers l'armée de métier* (*Towards a Professional Army*) in 1934 and *La France et son armée* (*France and its Army*) in 1938. He stressed the importance of a building a mechanized army with specialized tank divisions manned by professional soldiers, instead of static defences such as the Maginot Line. De Gaulle's ideas, like those of Fuller, Guderian and Tukhachevski, were not always well received. Pétain and other French politicians rejected them, largely because they questioned the political reliability of the professional army which he proposed.

Experiments with the use of tanks in roles other than infantry support finally began in 1932, initiated by the cavalry divisions. These experiments eventually gave rise to three light mechanized divisions, the so-called Divisions Légères Mécaniques – a fourth was formed in May 1940 – each consisting of 220 tanks and an infantry brigade. One should not be misled by the term 'Légères Mecaniques': this did not in fact mean 'light mechanized', but 'mobile'. These divisions, which were balanced in terms of structure, were in danger of being deployed in a very dispersed manner, based on old cavalry doctrine, in a wide screen in front of other units. After the German invasion of Poland in September 1939, the success of which was mainly delivered by the German tanks and their cooperation with the Luftwaffe, the French hastily began to form four new tank divisions with heavier tanks and less infantry. These tank divisions had a different role to the German ones: they aimed to create a breach in the front, allowing conventional units to advance. They thus effectively retained their traditional role as infantry support units, since the French were not convinced that tank formations could advance deep into the enemy's rear to disrupt lines of communication and cripple C3I capability.

The tank doctrine developed thus greatly differed from the German one, and a strict separation was maintained between infantry and cavalry tanks.

We will now look at the various French tanks, because this gives an interesting picture of the position of the French Army in 1940.

Infantry tanks
Char B1

The Char B1 was originally developed in the 1920s as the Char de Bataille based on a concept by Estienne. Early in 1927, it was decided to build three prototypes of the Char B, based on four earlier experimental vehicles. In 1930 the prototypes were thoroughly modified to meet new specifications, and in April 1934 the first order was placed for seven tanks of the Char B1 type.

The Char B1 was produced by several companies: Renault (182), AMX (47), FCM (72), FAMH (70) and Schneider (32). The tank was very expensive to build, each costing around 1,500,000 francs. This put it at the heart of the debate between two French schools of thought: one wanted to build a limited number of very powerful heavy tanks, the other a large number of light and cheap tanks. The debate eventually resulted in there being never enough tanks of either category, to the despair of de Gaulle, for example, who wanted more medium-weight Char D2s to be built; these cost a third of the Char B1, but were armed with the same 47mm gun. In appearance,

The prototype of the Char B had its first test drive in 1929. The silhouette of the tank had already acquired its definitive lines; only a few details on the superstructure would change.

the Char B1 reflected the fact that the development of this tank had begun in the 1920s; like the first British Mark 1 tanks of the First World War, it had large tracks around the hull and armour plates protecting the suspension. And like all other tanks of that period, the armour was not welded or cast, but riveted. The similarities with the Mark 1 came from the fact that the Char B1 was specifically developed as an offensive weapon, a tank that had to force a breakthrough through heavy defences and be able to clear trenches. The tank was seen as a mechanized and armoured gun that would destroy enemy infantry and artillery positions with its 75mm howitzer and was in fact built around this weapon. At that time, the French Army still assumed that capturing key positions at the front could be decisive, and it considered itself lucky to be the only army in the world to have a sufficiently large number of well-armoured heavy tanks. Speed was of secondary importance for this tank, because after the front had been pierced, the infantry, with or without support from their own infantry tanks, would take over the initiative.

When it became clear in the 1930s that enemy tank attacks also had to be repelled, it was too late for a completely new design. A standard APX-1 turret, also found on the Char D2, was placed on top of the hull. Whether this was the reason that the Char B1 performed less impressively on the battlefield than its armour and armament suggested it would, is unclear. In 1940, the majority of the Char B1s' combat losses were due to German artillery and anti-tank guns. In tank-to-tank combat, the Char B1 usually had the upper hand, as is clearly illustrated by a Char B1 named 'Eure', which on 16 May ambushed and knocked out PzKpfw III and IV tanks; it then managed to withdraw safely despite having been hit 140 times itself. The Char B1 had 40mm frontal and side armour, and the APX 1 turret was equipped with a 47mm gun. This gun had shells that could only penetrate 25mm of armour. The 75mm howitzer was placed on the right side and two 7.5mm machine guns, one in the hull and one in the turret, completed the set. The 75mm howitzer could fire both armour-piercing and high-explosive shells, but could only be turned to a limited extent (one degree to left and right gave a field of fire 18m wide at a distance of 500m) – in other words, the driver had to aim the tank as a whole at the target. The 75mm gun had its own loader, and the other two crew members were the radio operator (the E35 radio used Morse code) and the commander, who had to aim and load the 47mm gun. In the rear wall of the fighting compartment was a door leading to the engine compartment, where a team of three (!) mechanics kept the 272hp engine running. They also supported the crew in battle. The engine provided a speed of 28kph. A total of thirty-four Char B1s were built between 1935 and 1937. The Char B1 was succeeded by the Char B1 bis with thicker armour (60mm maximum, 55mm on the sides) and an APX 4 turret with a new 47mm gun providing real anti-tank capacity. Between April 1937 and June 1940 a total of 369 Char B1 bis tanks were delivered out of a total order for 1,144.

The Char B1 was placed in the tank divisions of the infantry, the *Divisions Cuirassées de Réserve*, which specialized in breaking through fortified enemy positions. After the breakthrough, the mobile phase of the attack was taken over by the infantry, supported by their infantry tanks or the tank divisions of the cavalry, which were equipped with the SOMUA S 35.

Light infantry tanks

In addition to the heavy Char B, the infantry increasingly had access to light tanks in the 1930s. France was the only country to have retained from the First World War a large number of light infantry tanks of the Renault FT 17 type, which was at the time considered to be very modern. Due to this large stock of tanks and the relative lack of international tension in the 1920s, there was little urgency in the search for replacement types. In 1926, the infantry drew up a plan for the development of a cheap *char d'accompagnement*, a light tank for direct infantry support. This 'Plan 1926' eventually led to the production of the Char D1 in 1932.

At that time, however, it had already become clear that this tank was neither light nor cheap and was also urgently needed as a battle tank – to combat enemy tanks – because no other modern type was available. The niche for a light infantry tank therefore still existed, and the FT 17s had become completely outdated due to rapid technical developments in the period. However, the infantry hesitated to start a new project, because many officers in the high command felt that the limited budget could be better spent on heavy tanks. After all, the introduction of large numbers of light anti-tank guns had made light tanks far too vulnerable, which threatened to render the then current tactic of overwhelming the enemy defence with a 'swarm' of light tanks – people spoke of 'saturating the battlefield' – ineffective.

The management of Hotchkiss, an automobile and arms manufacturer, believed they had the technical solution to the vulnerability of light tanks and in 1933 presented, unsolicited, a full-size plaster model of the tank they were to develop. It was light (estimated at 5–6 tons) but, thanks to the use of cast steel, was nevertheless very resistant to anti-tank guns of the calibre that was common at the time. Cast steel offered several advantages: it allowed the use of rounded shapes that could both encompass the interior of the tank as efficiently as possible and deflect incoming shells; complicated structures could be cast in one process, which saved hours of work and limited the number of gaps, as protection against the feared use of poison gas.

The proposal was very well received, partly because Hotchkiss was already working on a soft-iron prototype. Hotchkiss based its design on the idea that the French Army did not aim to carry out deep strategic penetrations into the enemy's rear; offensive actions would instead have limited depth, and the infantry would play a leading role. The Hotchkiss tank had to support the infantry by concentrating on weak spots (compare the German *Schwerpunkt*), neutralizing anti-tank guns and machine-gun nests, widening the breakthrough and further accompanying the infantry in rolling up defences. Following Hotchkiss' initiative, the army asked all of French heavy industry to come up with alternative proposals, the so-called 'Plan 33'. No fewer than sixteen other manufacturers responded and were invited to produce a prototype at the state's expense. In addition to Hotchkiss, these were: Atelier de Construction de Puteaux (APX), Compagnie Général de Construction des Locomotives ('Batignolles-Châtillon'), Berliet, Citroën, Delaunay-Belleville, Forges et Chantiers de la Méditerranée (FCM), Laffly, Lorraine Dietrich, Renault, Saint Nazaire-Penhoët, SERAM, Société d'Outillage Mécanique et d'Usinage d'Artillerie (SOMUA) and Willeme. All were given an assignment for one prototype. Five actually presented a vehicle: besides that of Hotchkiss itself, the Char Batignolles-Châtillon, the APX 35 of the state-owned APX, the FCM 36 and the Renault R 35 of France's most important tank manufacturer.

Hotchkiss

The original Hotchkiss prototype was presented to the Commission d'Expérience du Matériel Automobile (CEMA) in Vincennes on 18 January 1935. In June 1935 the type was approved in principle, on condition that it still met the new specifications. However, there was criticism of its poor visibility and the very narrow seat for the commander/gunner. A third prototype was tested and subsequently accepted for mass production as the Char léger modèle 1935 H. In November an initial order for 200 vehicles was placed. The first series was again extensively tested, during which major mechanical problems came to light: the transmission, the brakes and the steering were too weak, which resulted in increased wear, and the tank proved difficult to steer, particularly off-road. The H 35 in fact proved to be a danger to the men it was supposed to accompany. The infantry therefore decided only to purchase the first hundred vehicles to equip two tank battalion, the 13th and 38th Bataillon de Chars de Combat, and to refrain from further orders, in favour of the R 35 and FCM 36 already ordered. The money thus saved was transferred to the purchase of the Char B1. The normal consequence of that decision would have been to halt production, but that was unthinkable,

The Hotchkiss prototype from 1935, a 6-ton infantry support tank with two crew members. In practice, the tank did not perform well, and the infantry preferred the Renault R 35 and the FCM 36. The tanks already produced were divided over ten cavalry squadrons and four tank battalions of the infantry.

for political reasons. The remaining 300 vehicles were therefore forced on the cavalry, which had seen its original wish to equip its armoured divisions with the medium-heavy SOMUA S 35 rejected. The combination of the Hotchkiss H 35 and the SOMUA S 35 proved in practice not to be a happy one, due to the difference in speed between the two tank types.

The Hotchkiss H 35 was a small vehicle, weighing 11,370kg unladen. It had a crew of two. The driver sat in the front right and the commander sat in the fighting compartment and manned the standard APX turret. The turret was armed with a short 37mm gun, the shells of which could penetrate a maximum of 23mm of steel, so it posed a threat only to the most lightly armoured German vehicles of the time and then only at very close range. At the start of the Second World War, 640 Hotchkiss tanks had been delivered, according to the army, including 232 to ten cavalry squadrons and 180 to four infantry tank battalions. The rest were in North Africa, in storage or in reserve. That year, it was decided to use the greater part of light tank production capacity to build one type, the thoroughly modified Hotchkiss Char léger model 1935 H modifié 39: this tank was reasonably armoured and armed – no less so than the German medium Panzerkampfwagen III – not too expensive to make and, unlike all other French light tanks produced at that time, sufficiently mobile to be of use in the many armoured divisions that were to be deployed for the planned great summer offensive against Germany in 1941; the French intended to deploy 4,000 Hotchkiss tanks for this purpose. In order to enable sufficient mass production, British and Portuguese heavy industry would also be involved in the production of the cast armour parts; these could of course only be made in foundries. Production was planned to be increased to 300 per month from October 1940, and even to 500 from March 1941. These numbers were never close to being achieved, and total production remained at around 1,200.

Renault R 35

As we have seen, one of the other manufacturers that responded to the request was Renault, who feared losing their place as France's leading tank manufacturer. Renault did its utmost to beat Hotchkiss to it and was the first to finish its prototype on 20 December 1934. Renault's haste had a downside, however: the R 35 was too advanced in development to be adapted to new specifications that provided for an increase in weight to 9 tons and an armour thickness of 40mm. After the necessary adjustments, the first production example of the R 35 was delivered on 4 June 1936, and 380 R 35s were produced that same year. Production got off to a good start in the years that followed and, because of the threat of war, it was decided to increase it to 40–50 per month by 1939, then gradually increase it to 60 per month in 1940. This level of production was to be maintained as long as the war lasted. The number of vehicles officially ordered grew to 2,300, but in reality only 1,685 were produced, some of which were intended for export. This made the type the commonest French tank of its time. The R 35 was a typical infantry tank with 43mm armour, a 37mm cannon and a 7.5mm machine gun. Its speed was 20kph, sufficient to keep up with the infantry, but far too slow for mobile warfare.

The R 35 would have been most effective in the aforementioned independent tank battalion as part of an infantry division, in which the tank would support the infantry, hence the name 'Char d'Accompagnement'. Although the R 35 was a typical infantry support tank, in practice little training was conducted with the infantry. In addition, the tank battalions in

The Hotchkiss's counterpart, the Renault R 35, also weighed 6 tons and had a crew of two. The R 35 was an infantry support tank, hence the name 'Char d'Accompagnement'. In practice, however, little training was done with the infantry.

which the R 35 was placed were ultimately not distributed among the infantry divisions; instead, two or three battalions together formed a strategic reserve at army level in so-called *Groupements de Bataillons de Chars*. These groups were pure tank units and did not have supporting infantry or artillery, but the low speed and weak armament of the R 35s made them unsuitable as a mobile reserve to intercept and destroy enemy armoured units that had broken through. The lack of a radio also made it almost impossible to maintain control and coordination during mobile or non-mobile confrontations.

In addition, the idea of equipping each infantry division with its own tank battalion would have required a total of 5,000 light tanks. Although progress had already been made in that direction by June 1940, training was too far behind to actually man all those tanks. In addition, some of the crews were needed to man the heavier types in the armoured divisions. On 10 May 1940, at the start of *Fall Gelb* (the German invasion), there were twenty-one battalions with a total of almost 900 R 35 tanks available. In addition, two battalions were equipped with FCM 36 tanks.

FCM 36

One of the other manufacturers that accepted the invitation to design a prototype as part of Plan 33 was the shipyard Forges et Chantiers de la Méditerranée (FCM) in Toulon, which had produced the Char 2C in 1921. In naval construction, a lot of experience had already been gained with more modern techniques such as electromechanical welding of armoured steel plates, yielding a lighter construction with the same strength. FCM would use this method for its tank, the FCM 36. The hull and turret were almost entirely welded, except for the roof of the engine compartment, which had to remain completely removable to allow maintenance.

A full-size wooden model was delivered as early as March 1934 and was approved. During tests by the Commission d'Infanterie in 1935, many imperfections came to light, which led to multiple adjustments. Finally, after a heavier engine was introduced and the armour was improved, the aforementioned commission declared that this was the best tank, although improvements were still possible. As early as May 1935, when the Rhineland crisis reached its climax, an order had been placed for 100 units of the Char léger Modèle 1936 FCM or FCM 36, at a cost of 450,000 francs each. The tank had 40mm of armour, a 37mm gun and a 7.5mm machine gun, and a speed of 24kph. However, it was not the only tank purchased by the infantry, since two from competitors had already been purchased on a large scale: the Hotchkiss H 35 and the Renault R 35. These were considerably cheaper than the FCM 36, which was nevertheless purchased because the comapny's plant in Toulon was inaccessible to the German bombers of the time

In addition, it was believed that this tank's advanced design could show France the way to the future. After the first order of a total of 300 units, FCM declared that the costs had been wrongly calculated and that the tanks would cost 900,000 francs (!) each, double the original price. This would make the FCM 36 no less than three times as expensive as other light tanks. Ultimately it was decided to abandon further production, which was not all that inconvenient for FCM as they were busy with the production of Char B1 bis and did not believe they could deliver any FCM 36s before September 1940.

The FCM 36 was also a pure infantry support tank for which off-road performance was most relevant: it could cross a 2m ditch, drive through water 1m deep, negotiate an obstacle 70cm high and climb a 30° hill. The weak point of all French tanks of that period, however, was the limitations of the APX1 tank turret: the tank commander could not put his head out of the turret, but had to sit on its rear hatch to gain an unobstructed view. In addition, he also acted as gunner and loader of the standard armament of the light infantry tanks: the 7.5mm machine gun and the 37mm gun. Since the shells of this gun could only penetrate 23mm armour, their effect against enemy tanks was very limited. Like the R 35, the FCM 36 had no radio on board, so communication between tanks was not possible.

The FCM 36 was used to equip two battalions: as the only French battalions equipped with light tanks, these two were designated Bataillon de Chars Légers or BCL: the 4th Bataillon de Chars Légers was created in March and the 7th BCL in April 1939. Each battalion had a strength of thirty-nine tanks, made up of three companies of thirteen, plus a parts company.

Cavalry

After the experiments of the early 1930s, the cavalry prepared for its new role. On 26 June 1934, the new specifications for an Automitrailleuse de

Combat (AMC) were announced. 'Automitrailleuse' ('self-propelled machine gun vehicle' or armoured car) was a euphemism used to hide the fact that it was in fact a tank, since by law only infantry were allowed to use tanks. The specifications demanded 40mm armour, which would make the tank invulnerable to most anti-tank guns of the time, a 47mm gun and at least one machine gun. The tank had to have a speed of 30kph on the road, a range of 250km and carry a crew of three. As so often happened in France when purchasing weapons, there had already been prior contacts between the army and industry. Renault was already developing a type in response to the first specifications: the AMC 34. However, after previous experiences with the AMR 33, the cavalry had serious doubts that this would result in an useable vehicle. They were therefore happy to accept a plan from a subsidiary of the Schneider company, which had produced one of the first tanks in the First World War – the Société d'Outillage Mécanique et d'Usinage d 'Artillerie or SOMUA, based in Saint-Ouen – to build a prototype that would use cast steel.

On 17 May 1934, the design was submitted to the Service de l'Armement, and the contract was signed two months later. First, two wooden models were made on a scale of 1:5, and then in April 1935 a prototype, the AC 3 ('AC' stands for Automitrailleuse de Cavalerie) was completed. While the AC 3 was being tested and drastically modified in a number of ways (it was too slow and its engine was extremely noisy), SOMUA developed the AC 3 further into the AC 4. This type was extensively tested in 1936 and 1937, eventually leading to its approval in early 1938. However, political realities overtook the tests: due to the Rhineland crisis, the international situation had deteriorated so much that on 25 March 1936 the AC 4 was accepted as the standard medium tank of the cavalry with the official name Automitrailleuse de Combat modèle 1935 S (or AMC 1935 S). Informally, the vehicle was usually called the SOMUA S 35. At the same time, an initial order was placed for fifty tanks.

The cavalry had originally planned a total purchase of 600 tanks in order to equip its three armoured divisions almost completely with the type. These were the previously described Divisions Légères Mécaniques (DLM): they formed the most heavily armed units on the Allied side, partly due to a large contingent of SOMUA S 35s. Due to budget cuts, the number had to be limited to 300; the shortfall would be made up with 300 Hotchkiss H 35s, the light infantry tank that had been rejected by the infantry. However, this tank was slow and lightly armed and was a poor companion to the SOMUA S 35. Just before the war, a third order was placed for 200 tanks, both to set up new armoured divisions for a major offensive against Germany in 1941, and to replace the slow Hotchkiss H 35s of the original series. It was hoped to have seven DLMs, with a total of 800 tanks operational in the summer of 1941; the original three divisions with a nominal 160 tanks each and four armoured divisions yet to be set up with a core of eighty SOMUA tanks supplemented with the Hotchkisses.

The SOMUA S 35 had a maximum armour thickness of 47mm at the front, 40mm at the sides and 35mm in the rear; it could overcome obstacles 50cm high, climb a hill of 35° and drive through water 1m deep. It had a so-called 'one-and-a-half-man turret' as opposed to the 'one-man turret' of most French tanks. The commander, as with infantry tanks, had several tasks: he was also gunner and loader of the 47mm gun and the coaxial 7.5mm machine gun,

The SOMUA S 35 was reasonably mobile and was better armoured and armed than its main opponent, the German PzKpfw III. However, it was mechanically unreliable and there was no space for communication equipment inside, which made direction and coordination between multiple tanks virtually impossible. However, after the necessary modifications, the Germans would make grateful use of captured SOMUA tanks.

which could swivel ten degrees to the left and right when disconnected from the gun. The radio operator could support the commander. The problem was that, here too, there was no radio for communication with the other tanks of the battalion. However, platoon and squadron commanders did have a radio for contact with higher echelons. One of the reasons for the missing radio was that its originally planned position was on the left rear inside of the turret. There was still room there, but it was to aborb the recoil of the gun: a radio in that position would be shattered by the first shot. In the spring of 1940, the tanks of some units were equipped with an intercom system. The commander had a view of the surrounding area via a 40mm-thick command cupola that could rotate independently and was equipped with four viewing slits. This did not provide a truly unobstructed view, and there was no hatch in the command cupola so that a commander could glance quickly outside. In practice, the commander sat on the rear hatch, the top of which could fold inwards. In this position, he could observe well, but he was also extremely vulnerable and had to lower himself back in to operate the gun.

The SOMUA S 35 proved to be an excellent match for the original cavalry specifications: the tank was reasonably mobile and was better armoured and armed than its likely opponents, the Soviet BT-7 and the German PzKpfw III. For example, the German tank was vulnerable to the SA 35's

The French Army also had so-called 'automitrailleuses' such as this AMR Citroën Kégresse P 28: a half-track equipped with a machine gun. This example belongs to the 7th Chasseurs of the 3rd Groupement de Cavalerie.

An AMR Schneider Kegresse P 16 of the Groupe d' Escadrons de Reconnaissance (GER) during manoeuvres in January 1940. The AMR P 16 was lightly armed and was designed for fast reconnaissance missions. The usefulness of the roller at the front of the vehicle is clearly evident.

gun at a distance of up to 800m, while it had to approach to within 300m in order to penetrate the side armour of the SOMUA S 35. For this reason, the SOMUA S 35 is often called the 'best tank of the thirties'. Operationally, its main drawback was its mechanical unreliability. Although this was no more severe than that of other French tanks, its need for regular maintenance was too great, given its special task in mobile warfare. In addition, its high cost meant that the type could only be purchased in very small quantities. The infantry was also interested in the SOMUA S 35: in 1938, General Pierre Héring proposed to purchase the SOMUA S 35 as a replacement for the Char B and to use it for an accelerated build-up of its Divisions Cuirassées. However, budgetary restrictions and the imminent threat of war ultimately forced the French to maintain production of the heavy Char B1.

Laffly 35

The Laffly S 15 was a family of military all-terrain vehicles from the French manufacturer Laffly that shared the same six-wheel chassis. The Laffly S 15T was a light artillery tractor that was used to tow light field guns such as the modernized 75mm gun or the 105mm howitzer. An interesting variant was the Laffly S 15TOE (théâtre d'opérations extérieure), a reconnaissance vehicle that was developed and built for service in France's African colonies. The S 15 chassis was retained, but an armoured cab protected the engine and crew, and the vehicle had a small turret with a single machine gun.

SOMUA MCG

The SOMUA MCG was an artillery tractor and recovery vehicle used to tow medium-weight artillery pieces such as the 155mm 1917 howitzer and the 105mm 1936 field gun, together with their dedicated ammunition trailers. In total, 345 of these vehicles were produced, 264 up to September 1939 and another 81 by the end of May 1940. There was also a special version equipped with a crane and intended to recover stranded tanks; a total of 440 examples of this type were produced.

The Germans captured more than 2,000 of these SOMUA MCG half-tracks. This example is equipped with roof armour and a Nebelwerfer installation. The weak point of these vehicles was their unprotected front wheels.

Unic P 107

During the 1920s and 1930s, Citroën developed a long line of semi-tracked vehicles based on the Kégresse patent. In 1934, the company introduced its latest and most powerful model, the P 107, as a successor to the Citroën-Kégresse P 17. However, Citroën went bankrupt before mass production could begin, and the new owner, Michelin, decided to focus on the civilian market. Unic was therefore able to take over the Kégresse patent and production of the P 107.

Two major variants of the P 107 were adopted by the French Army: a light tractor for the 75mm field gun and the short 105mm artillery gun, and a transport vehicle for engineer units. More than 2,000 examples were in service in 1940.

From the above sketch, it appears that France, too, finally came to terms with the potential of the tank. The most important difference from Germany and the Soviet Union was that the French saw no offensive role for the tank other than breaking through the fortifications on a relatively static front; this breakthrough would then be followed up by the infantry with infantry support tanks and cavalry tanks which would advance a short distance into the enemy's rear.

Chapter 7

The New German Armour in Action

The Austrian Anschluss

German armour would be put to the test for the first time in 1938 during the Anschluss with Austria, in which Guderian would play a central role. On 10 March 1938 he was summoned to General Beck, still Generalstabschef des Heeres (Chief of the Army's General Staff) to learn that Hitler intended to bring Austria into the Reich. A number of units could expect to receive marching orders in the following days.

Beck initially intended to have Guderian lead his old 2nd Panzer Division, but Guderian rightly noted that it already had a capable commander in the person of General Veiel. Guderian therefore proposed to place some other mechanized units, such as those of the Waffen SS Division SS-Leibstandarte Adolf Hitler, under his command within the XVI Army Corps. Beck agreed to this, and in the evening Guderian began preparations for the action. The 2nd Panzer Division and the SS-Leibstandarte Adolf Hitler Division under Sepp Dietrich were to assemble near Passau. During the evening Guderian learned that the entire operation would be under the command of Colonel General Bock and that infantry units would advance south of Passau via the Inn River and through Tyrol.

Further preparations were not without complications: some of the staff were on an exercise with General Veiel near Trier and had to return in a hurry; some of the tanks were not deployable; maps of Austria were unavailable; most important of all, they only managed to find 10 per cent of the designated tanker trucks, and even an unauthorized requisition of trucks in the wider area could not make up the deficit. In addition, it turned out the following day that fuel consumption by the tanks and other vehicles for the 375km between Würzberg and Passau had been grossly underestimated, so that by the evening of 11 March only half of the division had trickled into Passau.

The divisional staff and General Veiel had also arrived in the meantime, but fuel and maps of Austria were still lacking; as an emergency measure a Baedeker travel guide was used. Although there was a military fuel depot in Passau, this was only supposed to be used in the event of mobilization. The officers responsible for the depot had not been informed of the arrival of a complete division and could not be reached in the middle of the night. The night watchman refused Guderian's units access and only allowed them in after pressure was applied. Although fuel was now available, they still had no mobile supply units. The Mayor of Passau did manage to requisition some trucks, but this did not really solve the problem: the 2nd Panzer Division could only hope that the petrol stations along the road to Vienna would remain open!

The first tanks finally managed to cross the Austrian border on 12 March at 9am, instead of the agreed time of 8am. They were welcomed with open arms: hands were shaken, tanks were garlanded with flowers and the men were given food and drink from all sides. The first large city reached was Linz, where Guderian was invited to lunch by the local authorities. Himmler also showed up and said that Hitler planned to visit the city in the afternoon. This was no coincidence, since Linz was the Führer's birthplace, and he had a triumphant return to it in mind. Guderian was asked to make the necessary preparations, and towards evening Hitler arrived and addressed the assembled population from the balcony of the town hall. In his memoirs Guderian described how emotional Hitler was:

> When he addressed the enthusiastic crowd below, I stood next to him on the balcony of the Linz town hall and was able to observe him from close by. Tears were rolling down his cheeks and he was certainly not acting. It was perhaps the only time I saw him actually moved.

Guderian informed Hitler of his units' progress and at 9pm left for Sankt Pölten, which the reconnaissance units of the 2nd Panzer Division had already reached earlier that day. The Führer gave the order to continue the advance immediately, and Guderian's units reached Vienna at 1am on 13 March. There, all sorts of festivities celebrating the Anschluss had just ended, and the arrival of the first German units provoked a further wave of enthusiasm. In the presence of the commander of the Vienna district, General Stumpfl,

German units moving through Vienna, 1938. The Pak 37 anti-tank gun behind the truck would prove too weak to withstand modern tanks in the coming years.

the first German units marched behind an Austrian army band past the Opera. The next day was devoted to the preparation of a large joint military march on 15 March, in which Austrian and German units would march through Vienna together.

Evaluation of the panzer forces

Despite all the apparent success, the Panzer forces had learned a lot from the Anschluss. First of all, the supply system had failed completely, even under peaceful conditions. The fact that the German units had to fall back on the network of Austrian petrol stations speaks volumes. The lower octane content of Austrian petrol led to additional engine problems and higher fuel consumption. The wheeled vehicles had encounterd relatively few problems, but a considerable number of the tanks had broken down with mechanical problems. According to the *Erfahrungsberichte* (reports of experience), the division was 'hardly to be considered a fighting formation' upon arrival in Vienna. It was therefore not surprising that criticism of this new weapon was great and doubts were raised as to whether tanks were capable of being deployed for long periods and over great distances. Von Bock was also displeased with the decorations and flags that had been placed on the tanks at Guderian's initiative, an objection that he immediately dropped when he learned that the matter had been short-circuited with Hitler via Sepp Dietrich.

Guderian made a number of justified comments in response to the criticisms:

- The units were not prepared for such an action, since most had just started training at company level. The staff were practising intensively in the Trier region
- The OKW (armed forces high command) was also not prepared because everything had to be organized at short notice on Hitler's initiative; this meant that everyone had to improvise
- The distances that had to be covered within 48 hours were great: for the 2nd Panzer Division it was almost 700km, for the SS-Leibstandarte Adolf Hitler division, which was stationed in Berlin, it was almost 900km. Such distances were a challenge to any vehicles of that time, let alone tanks
- The weakness of maintenance facilities had already been noted during the manoeuvres of 1937 and hard work was being done to remedy it
- The problems in the area of fuel supply were genuine, and investigations had to be undertaken into how these could be remedied
- It turned out to be quite possible, as expected, to have more than one division travelling on the road system.

All in all, the Anschluss provided valuable information for future operations. Guderian and the new armoured units learned a lot from it.

The 2nd Panzer Division remained stationed in the Vienna district and began to incorporate Austrian recruits into its ranks in the autumn of 1938. The SS-Leibstandarte Adolf Hitler Division and the staff of XVI Army Corps returned to Berlin in April. Because there was room in Würzberg after the departure of the 2nd Panzer Division, a new division was established in the autumn of 1938, the 4th Panzer Division under General Reinhardt. The 4th Leichte Division and the 5th Panzer Division followed shortly thereafter. Guderian visited the units under his command in the meantime to build personal relationships with the officers and men, one of the basic conditions for mutual trust in war.

Parade of Panzer Regiment I and the SS-Leibstandarte Division on the Theaterplatz in Karlsbad on 10 October 1938. Hitler is in the foreground, with Guderian diagonally behind him.

In October 1938, Hitler and von Brauchitsch discussed the idea of appointing Guderian as General der Panzertruppen with authority over all motorized units, including tanks, motorized infantry and motorized cavalry. It was a kind of Inspektorat function, with the aim of supervising the development of the armoured divisions, the light mechanized infantry divisions and the mechanized cavalry. He would be authorized to conduct inspections and to draw up annual reports, but not to hold command, and he would have no authority over the creation and implementation of tactical manuals or over organizational and personnel matters.

Guderian rightly considered this role to be purely administrative and one in which he could exercise little power over the further development of mechanized units. He suspected malice on the part of the more conservative elements within the army hierarchy and refused the appointment. Even after pressure from the high command via Bodewin Keitel, a friend of Guderian from his time at the Kriegsakademie and the younger brother of Wilhelm Keitel, responsible for personnel matters, he continued to refuse it. In an attempt to persuade Guderian, Keitel confided to him that the idea had come from Hitler and that he could therefore simply not refuse. Guderian again refused, however, and Hitler summoned him to a personal interview, at which Guderian explained his refusal and emphasized his need to retain control over the development of the armoured units. This was the reason, he said, that he wanted his old job back. Hitler refused, but ordered Guderian to report to him personally if he was hindered in any way in the execution of his new task – which, Hitler emphasized, included building the entire mobile force. Guderian agreed on these terms and he was duly appointed General der Panzertruppen.

One of the points that Guderian was unhappy about was the unit he would lead in the event of war: a reserve infantry division corps. General Erich von Hoepner, a former cavalryman, was given command of XVI Corps. In fact, Guderian had little to worry about in terms of the further development of the German panzer force: by the end of 1938 there were five panzer divisions with the prospect of a sixth, and the four existing light divisions could be quickly converted into full panzer divisions if necessary, now that production of the PzKpfw III and IV had begun. In addition, the occupation of Czechoslovakia in 1938 provided an abundance of excellent equipment. As a result, if war came, the German panzer force could theoretically take on any of its neighbours, except the Soviet Union.

The Polish campaign

After the successful annexation of Czechoslovakia, Hitler turned his eyes to Poland, a country he saw as an artificial creation by the Treaty of Versailles, which had granted much of the centuries-old German territory of Memel and Weichsel (now Vistula) to the newly formed Polish state. Hitler wanted to reclaim this territory, but knew that he had to make concessions to do so, especially since Great Britain and France had committed themselves to the support of Poland in March 1939. In order to gain greater freedom of action, he turned to the Soviet Union, with which a non-aggression treaty, the Ribbentrop-Molotov Pact, was concluded in August 1939. Secret clauses of the Pact provided for Poland to be partitioned between the two countries.

On 22 August all generals were called to Obersalzberg to discuss this non-aggression pact, and the main lines for the Polish campaign were laid out.

Guderian was somewhat taken aback by these developments. He was busy preparing for the autumn manoeuvres of the mechanized units. On 22 August he was ordered to take command of the newly formed XIX Motorized Corps in Pomerania. This corps formed the northern spearhead of the German invasion force in Poland and consisted of the 3rd Panzer Division, the 2nd and 20th Infantry Divisions (both motorized) and the units of the Corps itself. The 3rd Panzer Division was also supplemented by the *Panzer-Lehr* (demonstration) *Abteilung*, equipped with PzKpfw III and IV tanks.

Among the units of the corps was the *Aufklärungs-Lehr-Abteilung* from Döbberitz-Krampnitz. From these two *Lehr* units one of the most famous German panzer divisions would later emerge, the *Panzer-Lehr Division*.

Guderian's XIX Corps was part of the Fourth Army under Colonel General von Kluge, which in turn was part of Army Group North under von Bock. To the north of Guderian's units were Grenz Schutz (border protection) units under General Kaupisch, and to the south, II Corps under General Strauss. The mission of Guderian's units was to push through the Polish Corridor as the spearhead of the northern wing and enter East Prussia, which had been separated from the rest of Germany by the terms of the Treaty of Versailles. Guderian's units were then to advance through East Prussia, turn south, and make contact with the units of Army Group South. The Polish Army was to be eliminated by means of a classic encirclement battle (*Kesselschlacht*). Kaupisch, on Guderian's left wing, was to advance towards Danzig, and Strauss, on his right wing, towards the Vistula.

Guderian's advance would take him through his native region in West Prussia and past his birthplace, Kulm, which had become part of Poland under the terms of Versailles. The strength of the Polish units in the Corridor was estimated at three infantry divisions and the Pomorska cavalry brigade, which appeared to have a limited number of Fiat-Ansaldo tanks. The Polish side of the border had been fortified, and a second line of defence was expected at the Brahne River. The focus of the attack on Poland was not in the north, but in the south, where Army Group South had to fight its way through Silesia to the Vistula south of Warsaw. Army Group South had XVI Corps at its disposal for this, Guderian's former unit, now under General Erich von Hoepner, consisting of two panzer divisions and two light mechanized divisions. The area through which they would pass was considered excellent terrain for the deployment of tanks. And not unimportantly, the Germans had gathered the Third Army under the command of General Küchler in East Prussia to support the attack on Poland from the north-east.

The plan assumed that the Poles would only lightly defend the western part of Poland and would withdraw east to the Vistula. The Germans aimed to use a large pincer movement to prevent the Polish units from reaching the Vistula, and then encircle them. Within the Wehrmacht there was, however, great reluctance to launch the Polish campaign, as it was realized that, given the British and French guarantees of Poland's sovereignty, it could easily lead to a war on two fronts. The significant section of the officer corps with experience of the First World War could well imagine what the consequences might be. Guderian viewed the coming conflict with anxiety for another reason: his eldest son was serving as an adjutant in Panzer Regiment 35, while his youngest son was in the reconnaissance battalion of the 3rd Panzer Division, one of the units Guderian himself commanded.

The attack itself was originally planned for 26 August, but was postponed because there was still movement on the diplomatic front. On 31 August, the German units were again put on alert, and at 4.45am on 1 September, Guderian's units began to cross the Polish border in thick fog. Guderian advanced at the head of the 3rd Panzer Division and, almost immediately, nearly fell victim to his own artillery, which had opened fire against orders: the first shell landed 50m in front of Guderian's command vehicle, the second 50m behind it. Guderian ordered the driver to turn the half-track around, but this awkward manoeuvre caused the vehicle to end up in a ditch. With the steering mechanism bent, Guderian had no choice but to go to the corps command post, speak to the overly enthusiastic gunners and requisition a new vehicle.

The first serious fighting took place north of Zempelburg and around Gross-Klonia, land once owned by Guderian's great-grandfather, where the fog suddenly lifted and the leading tanks found themselves facing Polish defensive positions. The Polish anti-tank guns scored many hits and the Germans suffered their first casualties. The forward units of the 6th Panzer Regiment reached the Brahne later in the day and halted there, against corps orders, because it was assumed that they would not be able to force a crossing that day. When Guderian arrived to investigate the reason for the halt, he was approached by a young lieutenant who declared that the Polish units on the other side of the river were weak and that his men had just managed to save the bridge at Hammermühle, which had been set on fire by the Poles. Guderian rushed to the scene to find that tanks of the 6th Panzer Regiment and the infantry were firing randomly at well-entrenched Polish units on the other side. Guderian immediately ordered the 'idiotic' firing to cease, and a unit was sent across the river in rubber boats to reconnoitre the Polish defences. He then sent the tanks across the bridge, and a short time later, the Germans succeeded in establishing a bridgehead on the other side. The 3rd *Aufklärungsabteilung* was sent forward towards the Weichsel to assess the size of the Polish reserves in the rear and by 6pm all relevant units were on the east bank of the Brahne.

The following night, lack of combat experience led to considerable panic among some German units. For example, shortly after midnight, the 2nd Infantry Division (motorized) reported that, after having taken Konitz earlier that day, they had now been forced to withdraw under pressure from the Polish cavalry. Guderian told the commander that he had to hold out and rushed to division headquarters early in the morning. He took over command for a short time and led the division until it successfully crossed

German panzer troops during a rest in Poland.

the Kamionka River north of Gross-Klonia. The division then retained the initiative and was able to continue the attack on its own. This was typical of Guderian, who here displayed the characteristics common to many German commanders in the Second World War: like a kind of human dynamo, he raced across the battlefield, was always where the fighting was fiercest, where the *Schwerpunkt* was, sometimes leading the attack himself and personally exposing himself to danger. According to Guderian, mobility was crucial, and he drove his units forward. As a result, the forward units sometimes ran out of fuel, as on the second day of the advance: the logistical problems that had arisen during the Anschluss would continue to plague the Germans throughout the war.

Guderian's units advanced rapidly in the following days, managing to encircle several Polish formations and destroy artillery units, and witnessing the dramatic attack with swords and lances of the Polish Pomorska cavalry brigade. On 3 September, Guderian visited the 23rd Infantry Division and 3rd Panzer Division and had the chance to briefly meet his son Kurt and see the towers of his birthplace Kulm on the other side of the Vistula. By 4 September, Polish units to the west of the Vistula were completely surrounded, and the 3rd Panzer Division was eliminating the last pockets of resistance. Guderian's units could now receive new orders. The first days of fighting had cost 150 dead and 700 wounded, including a disproportionate number of officers, because they had fought at the front with their men. Among the fallen were the sons of General Adam, State Secretary von Weizsäcker and Colonel Freiherr von Funk. In addition, not everything had gone to plan: some of the Polish units had escaped encirclement and withdrawn to the north to the defensive lines of the Hel Peninsula, where they would continue to offer resistance until Poland surrendered.

Hitler and Himmler, accompanied by a then unknown infantry officer named Rommel, commander of the Führer-Begleit-Bataillon, paid the units a visit on 5 September and drove with Guderian along the entire route where fighting had taken place. Hitler expressed his surprise at the small number of casualties on both sides, which Guderian explained by the mobile nature of the battle and the effectiveness of the tanks. The tanks' work was also evident from the degree of destruction on the battlefield; Hitler had assumed that this had been the work of the Luftwaffe and was

Guderian, like other high-ranking German officers, was always at the front with his combat units. Here he is consulting with men from a regiment of the 10th Panzer Division.

surprised to learn that Guderian's tanks were responsible. In response to Hitler's questions about the experiences of the past few days, Guderian replied that accelerated delivery of the PzKpfw III and IV to the combat units and an increase in production were of great importance. The speed of these tanks was good, but their armour, especially at the front, had to be reinforced. In addition, the armour-piercing capacity of the guns had to be improved, which would require a longer barrel and heavier shells. The same applied to the anti-tank guns.

There was a short pause in the campaign in the following days, during which the German units regrouped. XIX Corps was reinforced by the newly formed and not yet fully operational 10th Panzer Division, which had supported the north-eastern attack from East Prussia. Guderian used this pause to go hunting in the forests of the castle where Napoleon had established his headquarters during the campaigns of 1807 and 1812. The second phase of the campaign would mainly have the character of a purge, because the fighting power of the Poles had been broken; their best units had been destroyed or were in retreat. The Polish air force was also no longer capable of action and had disappeared from the skies. In the meantime, there was considerable uncertainty about what to do next. The commander of Army Group North, von Bock, wanted to advance south-east in order to encircle the Polish units in a broad movement using his armoured and motorized units. The OKW rejected this suggestion on the basis of information that von Bock did not have: the contents of the secret clauses of the Ribbentrop-Molotov Pact and the agreed Soviet invasion from the east. Since no agreement had yet been made about the future demarcation lines, the OKW did not want German troops to advance too far to the east. Therefore, on 8 September, Army Group North was ordered to move south toward Warsaw.

In the meantime, the first units of Army Group South had broken through the Polish defences in western Poland and were advancing rapidly toward the Vistula south of Warsaw. The first attempts by the 4th Panzer Division to take Warsaw without further support, however, came to nothing. This, combined with considerable resistance at the Vissal and north-east of Warsaw, led the OKW to decide to follow von Bock's plan after all: the fact that the demarcation line agreed with Soviet forces was further east at the Bug River than originally planned gave the Germans scope to opt for this variant. The main lines of this phase of the Polish campaign were as follows. The Germans were to advance along four spearheads: the two middle ones were aimed at Warsaw with the aim of encircling the Polish units west of the capital, while the two outer spearheads were to carry out a large outflanking pincer movement to encircle the rest of the Polish Army east of Warsaw and make contact with each other at Brest-Litovsk.

This plan gave Guderian an unprecedented opportunity to put into practice his theory of deep strategic penetration into the enemy's rear. Guderian's units were initially deployed to cover the northern flank of the attack on Warsaw, but Guderian successfully argued that his armour would be better deployed in the large encircling movement than in support of this infantry attack. Guderian's XIX Corps was expanded with the 10th Panzer Division and the Lötzen Brigade, a newly formed unit that would garrison captured fortifications. The 2nd Infantry Division (motorized) was withdrawn from the front line and added to the reserve of the Army Group.

In the following days, Guderian was mainly to be found in the front line, often leading his troops himself. This was to compensate for the inexperience of many German units. For example, on the morning of 9 September Guderian received news that the infantry of the 10th Panzer Division had crossed the Narev River and taken the fortifications on the other side. The Lötzen Brigade would form the garrison there, and they successfully managed to cross the Narev in rubber boats on Guderian's orders to support the attack. During the day, however, the reports about the capture of the fortifications turned out to be based on a misunderstanding. Guderian crossed the Narev himself to find out what exactly was going on and discovered that all the tanks of the 10th Panzer Division were still on the other side. Guderian ordered his adjutant to ensure that these were transferred as quickly as possible and went to the front line himself. There the forward units were being relieved, a standard procedure during peacetime manoeuvres, but inappropriate during actual combat. Furthermore, no one knew exactly where the enemy was, no reconnaissance patrols had been sent out and the artillery observers had no idea what exactly was expected of them.

Guderian immediately halted the relief of the units, made sure that the artillery opened fire on the Polish fortifications and reconnoitred the terrain himself with the commander of the infantry regiment, discovering a German anti-tank battery which, thanks to the perseverance of its commander, had advanced far forward. Guderian was very disappointed with the course of events, and only after the tanks had been ferried across on pontoons under his leadership could the attack begin; the fortifications were then taken without much effort, but a day had almost been lost due to a lack of leadership and experience.

In the meantime, the engineers had started, on the basis of written orders from Guderian, to construct emergency bridges over the Narev for the 10th Panzer Division, and subsequently for the 3rd Panzer Division. These bridges had to be ready before midnight. To Guderian's astonishment, he found at 5am the next day that the engineers had dismantled the bridges on the basis of an order from the 20th Infantry Division, in order to rebuild them downstream for this division. Guderian was flabbergasted: the commander of the engineers had said nothing about Guderian's orders, and the end result was that the tanks and vehicles of the 3rd and 10th Panzer Divisions found themselves in a long queue on the northern bank and then had to be ferried across with difficulty using pontoons. It was not until late in the evening that a new bridge could be built. All this had frustrated the rapid deployment of Guderian's units.

Once across, the units led by Guderian advanced quickly towards the River Bug and Brest-Litovsk, which was reached on 14 September by the 1^(0th) Panzer Division and the 20th Infantry Division. On 15 September the ring around Brest-Litovsk was closed, but a surprise attack on the citadel failed because the Poles had blocked the entrance with a Renault tank. Brest-Litovsk was one of the oldest and strongest fortresses in Eastern Europe, and the Polish garrison put up tough resistance. An attack on 16 September also failed because the infantry regiment of the 10th Panzer Division, with Guderian in the front line, did not advance behind artillery fire, opened the attack too late and suffered heavy losses. The 3rd Panzer Division and the 2nd Division (motorized) meanwhile advanced further south to the east of Brest-Litovsk.

On 17 September the citadel was finally taken, at just the moment that the Polish garrison attempted a sortie to the west over one of the undamaged bridges on the Bug. The immediate cause was the arrival of Soviet units; the Poles preferred to become prisoners of war of the Germans – and not without reason: the Soviets would later murder more than 20,000 Polish officers at Katyn near Smolensk.

Since the Bug was to be the future demarcation line and Brest-Litovsk was on the east bank, the Germans were in the zone assigned to the Soviets and had to evacuate the area by 22 September. Thanks to the flexibility of the local Soviet commander, Guderian was given time to collect and remove all his wounded and all remaining German materiel. Thus ended the campaign,which appeared to the outside world to have been a complete success. However, Guderian's experiences showed him that there was still much room for improvement. While the various units of the XIX Corps returned to their respective garrisons in the following weeks, Guderian was reunited with his family in Berlin.

During the victory celebrations in the Reich Chancellery on 27 October, Hitler presented Guderian and twenty-three other officers with the Ritterkreuz, a newly created medal that replaced the earlier Prussian decoration Pour le Mérite. Like that order, it was awarded to senior officers for conspicuously successful leadership, or to lower-ranking officers for extraordinary bravery in the face of the enemy. But the award ceremony would not be Guderian's only meeting with Hitler that winter.

In order to strengthen the political faith of the officer corps, and in particular that of the generals, a series of lectures was organized in November, at which Goebbels, Göring and, on 23 November, Hitler himelf would eventually appear. To the surprise of Guderian and the other officers of the army and navy, the party line was set out as follows: the Luftwaffe generals, under the inspiring leadership of party stalwart Göring, were politically completely trustworthy; it was expected that the admirals would also follow Nazi ideology; but the party did not have unconditional confidence in the generals of the army. After the success of the Polish campaign, this hurt the army top brass. A debate broke out over how to deal with the situation, and von Reichenau suggested that Guderian discuss the officer corps' feelings about this lack of confidence with Hitler.

Evaluation of the Polish campaign

We have already seen that, due to the limited combat experience of the German units, Guderian had to intervene personally time and again. His experiences were not unique, and despite the great success of the Polish campaign, the OKH (Army High Command), on the basis of the *Erfahrungsberichte*, was very dissatisfied with the troops' performance. Among the recommendations from these *Erfahrungsberichte* about the Polish campaign were:

- officers of all ranks should be present at the front and not lead from behind
- front commanders tended to exaggerate their losses, the strength of the enemy and the difficulty of the terrain in their combat reports, and therefore had to learn to report more accurately
- German units were not sufficiently trained to carry out adequate reconnaissance missions.

In addition, the OKH found that:

- cooperation between the various army units was insufficient
- the infantry received too little support from its own heavy weapons and from the artillery
- infantry fire was uncoordinated
- the marching discipline of the infantry was poor
- troops adopted too linear a defence instead of the prescribed 'elastic' form
- the infantry was not always sufficiently equipped to engage in night battles or fight on difficult terrain
- traffic handling was not satisfactory.

A more worrying criticism was the lack of initiative among junior officers and NCOs, something that the army leadership placed great value on, as we have seen. That is why it was emphasized even more strongly from that moment on that these men should take the initiative themselves, within the limits of the orders issued. Officers were not always to wait for orders, but learn to rely on their own judgement. The reason for the disappointing performance of the (junior) officer corps was largely due to the large influx of new men after 1935. In line with this, an intensive training programme was set up under the title 'Ausbildung des Feldheeres' for the period October 1939 – May 1940 to remedy the weaknesses that had been observed. This programme included monthly evaluations of the combat effectiveness of the various divisions and corps. In addition, military doctrine and military standards were revised. The programme focused on the non-commissioned and junior officers and emphasized the importance of leadership by these men and by reserve officers, especially in combat situations. The role of the non-commissioned officer as leader, trainer and teacher was again highlighted. Finally, the need for improved discipline was emphasized.

Central activities in this training programme were:

- carrying out reconnaissance missions
- coordination and discipline in infantry fire
- cooperation between the various weapon units
- offensive and defensive tactics at dusk and during the night
- the transition from offensive actions to defensive positions.

This thorough programme recalls the autumn of 1917, when troops were trained in the new offensive doctrine of *Angriff im Stellungskrieg* (attack in static warfare). It also shows how critical the army was of its own performance and how well it responded to perceived weaknesses. This attitude was typical of the German Army throughout the Second World War, and they needed this preparation, because Hitler's urge to conquer was still undiminished: in late 1939 he turned his attention to the West.

Chapter 8

The Campaign in the West

Preparations

After the Polish campaign, Hitler and his inner circle deliberated on the next step. Before September 1939 the Wehrmacht had had no plan for an attack on France. Given the strength of the French Army and its allies, the Germans preferred not to consider a confrontation. Under pressure from Hitler, however, the Wehrmacht examined possible ways to take on the French and British armies. The first plan, drawn up in great haste – *Fall Gelb* – of October 1939, consisted of a limited offensive in the north followed by a Luftwaffe attack on Great Britain. The plan was considered uncreative and failed to convince anyone. In fact, it was a variant of the Schlieffen Plan, drawn up before 1914 – an attack through the lowlands of Belgium with the Channel coast as the first target – but it lacked the ambition of the original Schlieffen Plan, the destruction of the French Army. In fact, what the Wehrmacht envisaged was an almost nineteenth-century campaign, while the Germans had just proved successful in a rapid mobile war. The plan was also opposed by Hitler himself, who wanted to end the war in the West with one decisive offensive. However, the plan remained in force until mid-February 1940, when two careless officers with the complete plan in their briefcase crashed their plane on French soil. This cleared the way for a completely new and challenging variant, built around a southward flanking movement through the Ardennes as suggested by von Manstein.

In December 1939, von Brauchitsch and Halder had rejected von Manstein's proposal as being absurdly risky. By February 1940, however, the situation was completely different: a new, comprehensive plan was needed, and after much discussion, von Manstein's was accepted at the end of the month. This meant that the centre of gravity of the offensive was shifted from the Belgian plain to the Ardennes. The main force of the German armoured divisions would pass through the Ardennes in this southward movement.

The armoured units had to seize the Meuse crossings as quickly as possible, literally bypassing the Maginot Line, which ran to the south of the Ardennes, and then push on to the Channel coast. By means of this deep southward flanking movement, they would eventually be able to encircle a considerable part of the enemy forces, including many French and British armoured units who were expected to advance northwards through Belgium to confront the Germans. These would then be caught between the northern attack formation advancing through the Belgian lowlands and the armoured units in the south moving fast through the Ardennes.

It was a bold plan, which would result in stunning success for the Wehrmacht in May 1940. Guderian was only indirectly involved in its preparation, although von Manstein lacked experience with armoured units and sought advice from him, among others. Guderian could also draw on his experience in the early days of the First World War, when the units he was attached to had moved through the same area at high speed. Urgent preparations continued in the months that followed. During one of the war games devoted to the coming offensive, Chief of Staff Halder and Guderian disagreed about the speed of the offensive. Guderian favoured a rapid capture of the crossings over the Meuse and an immediate continuation of the advance. Halder had a more cautious approach, preferring to pause the advance after the capture of the crossings in order to regroup. It was clear that this subject would remain controversial: Guderian, of all people, was given the task of capturing the Meuse crossings and forcing a breakthrough.

Guderian remained commander of the XIV Corps, but the Corps was expanded with the 1st, 2nd and 10th Panzer Division and the SS Panzer Regiment Gross Deutschland. This large motorized regiment consisted of four battalions, which served in turn as Führer-Begleit-Bataillon (Führer escort battalion). In total, the Corps was an exceptionally strong force, consisting of twelve Panzer Abteilungen with a total of 800 tanks, three battalions of armoured cars, sixteen battalions of motorized infantry and sixteen batteries of anti-tank guns, anti-aircraft guns and field artillery. This force formed part of a larger unit, the Panzergruppe Kleist, which also included XL Corps and XIV Corps. Panzergruppe Kleist had a total of four armoured and three motorized divisions at its disposal. The northern flank of Panzergruppe Kleist would be protected by two other armoured divisions, the 5th and 7th Panzer Divisions, the latter under Rommel's command. The whole formed Army Group A. Army Group B under von Bock was responsible for the advance through the Belgian lowlands and had to face the French and British when they launched their expected advance towards Belgium and the Netherlands with the aim of halting the Germans along the Breda–Dinant line and preventing them from reaching the Channel coast. Von Bock had the 3rd, 4th and 9th Panzer Division at his disposal. Army Group C under the command of von Leeb was given a more or less static role in the Maginot Line area and had no armoured units.

However, it was questionable whether the German force that invaded the Netherlands, Belgium and France was eminently suited to mobile warfare with tanks and armoured vehicles. Of the ninety-three battle-ready divisions, only nine were armoured divisions, with a total of 2,439 tanks at their disposal. These units faced a French Army that was much better motorized, with a total of 3,254 tanks. Adding the British, Belgian and Dutch armour, the Germans' opponents had no fewer than 4,200 tanks at their disposal. And what the Germans lacked in quantity was not compensated for by the quality of their tanks. Compared to their French, British and even Belgian counterparts, the majority of the German tanks were of inferior quality. The majority of the French tanks, like the German ones, were housed in armoured units, which, on paper at least, were not inferior to the German ones. In addition, many French infantry units also had tanks as support. The French could afford this because of the large number of tanks they possessed. However, the structure of the French units and the doctrine they followed differed greatly from the Germans', as we have seen.

Nor did the Luftwaffe enjoy numerical superiority: it had 3,578 aircraft at its disposal, compared to the Allies' 4,469. In addition, the French had just received 500 aircraft of American make, including very good fighters that should have easily been a match for the Luftwaffe.

The advance to the Channel coast

At 5.35am on 10 May 1940, after months of 'Phoney War', the front in the West sprang to life. Thousands of German guns blasted Belgian, French and Dutch positions. Infantry units prepared to attack, and tanks and lorries started their engines. Thousands of bombers and fighters were on their way to targets in the West. The most dramatic action took place north of Army Group B's main line of attack, when the famous Belgian fortifications on the Meuse were taken by daring glider and commando attacks. Meanwhile, German paratroopers were landing near The Hague and at the bridges at Dordrecht and the Moerdijk to capture strategic positions. These spectacular attacks, combined with the advance of Army Group B on the Meuse, had their intended effect. The bulk of the French Army, supported by the British Expeditionary Force, advanced north. In response to the original plan of 1939, this would have been an extremely effective response by the French and British. In reality, their prompt reaction heralded a disaster of unprecedented scale. Undetected by Allied intelligence, seven of the nine armoured divisions of Army Group A were advancing along the winding roads of the Ardennes through Belgium and Luxembourg towards northern France. Their objective was to capture the bridges between Dinant and Sedan. If Army Group A were able to capture these bridges they could advance rapidly into the vacuum left by the French and British units that had moved north. These units would end up caught between Army Group B in Belgium and the tanks of Army Group A in northern France. The German plan of action was extremely daring by conventional military standards. If the French and British had left reserves of any size, the advance of Army Group A would have encountered serious resistance; they might themselves have become the victim of an encirclement.

In the event, things worked out wonderfully well for the Germans. On 13 May, the bridges at Dinant and Sedan were captured before the French and British were aware of any danger. The commanders of the German armoured divisions made the most of the opportunities offered to them, driving their units rapidly westwards into the rear of the French and British forces. In this way, they threw the Allies completely off balance and prevented them from building up a proper defence line. The feared counter-attack against the vast and vulnerable left flank of the German armoured divisions never got off the ground. Because the French had also moved all their reserves north, and their command centres had quickly lost contact with the situation on the ground, the Germans were able to retain full control. As a result, the bulk of the Anglo-French units were still moving north, while they were hemmed in by the units of Army Group A in the south.

Guderian in the western campaign

Guderian played a particularly active role during the western campaign, one which would bring him into conflict with his superiors on several occasions. His first battle objective was just north of the westernmost part of the Maginot Line at Sedan on the Meuse; he was tasked with moving through the difficult Ardennes as quickly as possible and capturing the river crossings. The offensive was to begin on 10 May, and it was expected that the German units would not encounter much resistance. This was exactly what happened. While all attention was focused on the German advance through the Netherlands and the Belgian lowlands, and the mobile reserves of the French Army and British Expeditionary Force were turning north, Guderian's units advanced through the Ardennes almost unnoticed and meeting little resistance. On the morning of 12 May, Guderian's forward units under Colonel Balck captured bridgeheads on the other side of the Semois, an eastern tributary of the Meuse, and that same evening, the commanders of the 1st and 10th Panzer Division reported that they had captured the town and fortress of Sedan.

But the capture of the city did not mean that the crossing of the Meuse was clear, because the French had blown up all the bridges before they surrendered.

Guderian on 10 May 1940 in his SdKfz 250 at the beginning of the deep advance of his panzer units.

The next morning, boats would have to be launched to capture bridgeheads on the other side of the Meuse, an operation that Guderian had carefully planned in the preceding weeks. Crucial to the success of the operation were the agreements he had made with Luftgruppe A, the Luftwaffe units that would support Army Group A. The officers of Luftgruppe A had participated in all discussions, and it had been made clear to them that the key to success was the continuous bombardment of the French batteries on the other side of the Meuse. The first successful crossing was made by units of Balck's regiment under the protection of Stukas. When Balck saw that the defenders in their substantial bunkers on the other side of the river were unsettled by the Stuka attack, he decided to seize the moment. He sent for the trucks with more boats and ordered his men to assemble them in front of the French, who were less than 50m away on the far bank, then sent his infantry across. The troops landed almost unopposed and immediately took up positions. When Guderian arrived shortly afterwards, Balck greeted him with the words 'Playing with boats on the Meuse is forbidden', a reference to a remark Guderian had made to one of his officers during an exercise, when he had the impression that the men were fooling around.

Once the Germans had gained a foothold, Guderian could give the order to build a bridge, enlarge the bridgehead and prepare to attempt a breakout. But von Kleist wondered whether it was wise to continue the advance so quickly. On the evening of 15 May, a heated discussion was held between Guderian and von Kleist. This was a fundamental difference of opinion that would recur in later years: should the tanks continue to advance quickly to maintain the element of surprise, or should regular halts be made to regroup and re-establish contact with the slower infantry? On this occasion Guderian managed to persuade von Kleist and he was allowed to continue his advance for another twenty-four hours. The next day, Guderian's units covered 60km and managed to make contact with the leading elements of Reinhardt's Panzer Corps. They had encountered tough resistance, but had the impression that this was local and improvised in nature. On this basis, Guderian made preparations the next morning to continue the advance. The first units had hardly started when he was ordered to stop immediately and report to von Kleist at an emergency airfield. There, an angry von Kleist accused Guderian of insubordination, whereupon Guderian tendered his resignation, which von Kleist accepted.

Guderian made sure that von Rundstedt heard about the incident as soon as possible and assumed that his contacts within the OKH would protect him. His assessment was correct, but from now on he was told to strictly follow orders from above. He was ordered by the OKH to carry out an 'offensive reconnaissance', an order that he would, not unexpectedly, interpret as broadly as possible. In order to prevent his orders from being intercepted by his superiors, he had a fixed telephone connection established between his Corps headquarter, which he was not allowed to move, and his forward staff headquarters. He then left for the front, where the offensive reconnaissance would take on remarkable dimensions. On 18 May, the 2nd Panzer Division reached Saint Quentin on the east side of the Somme battlefield, approximately at the point where the Germans had begun their attack on 21 March 1918. The next day, they were attacked by a whole tank

German troops cross the Meuse. Crossing this river and capturing the bridges enabled the Germans to carry out their large southward flanking movement.

Hanging out of his SdKfz 250, Guderian discusses the position of the 8th Panzer Division with Major General Kuntzen on 11 June 1940.

division under de Gaulle, which failed to halt the advance by a flanking attack. Two days later, at 9.30pm, the first units of 2nd Panzer Division reached Abbeville, where the River Somme flows into the sea; the trap was closed. It was a demonstration of the speed of the panzer units and the culmination of von Manstein's plan. In one major movement, Army Group A had encircled an area 200km long and 140km wide. North of the long and narrow corridor forced by Guderian's and Reinhardt's armoured units, large parts of the French, Belgian and British forces were located. It was the largest encirclement in military history: no fewer than 1,700,000 soldiers were caught between the hammer and the anvil of the German forces.

The next phase of the offensive was now underway, and Guderian's units turned north to capture the Channel ports and enclose Allied units in an increasingly tight circle. By 24 May, only the ports of Dunkirk and Ostend were available for evacuation of Allied units. On this day, however, Hitler issued an emphatic order to stop. He had two reasons for doing this: firstly, the corridor was too narrow and too weak, and Hitler wanted to strengthen and widen it by giving the infantry time to re-establish contact with the forward panzer divisions. Secondly, the countryside of Northern

Guderian, with his adjutant Riedel in the background, gives instructions to Oberst Leutnant Balck. Later, Balck would develop into an extremely capable commander of armoured units in the Soviet Union.

France did not seem suitable for panzer operations. He wished to maintain the strength of the panzers for the heavy fighting he expected in the south against the substantial remnants of the French Army, including Army Group Two. For thirty-six hours, to Guderian's frustration, the front came to a standstill; this gave the British time to strengthen their positions at Dunkirk

The beach at Dunkirk offers a dismal picture of abandoned British and French equipment.

A series of knocked-out French tanks marked the advance of the German armoured units. This is a Hotchkiss H 39 of the 14th Batallion de Chars de Combat, which was completely destroyed in the period 17-19 May in defence of the bridges over the Oise.

An FCM tank is being watched with interest by German soldiers. The soldier on the back of the turret is sitting on the hatch from which the tank commander usually observed the surroundings.

and then evacuate a large part of their army. Eventually, 370,000 British troops escaped via Dunkirk, as well as another 100,000 French troops who managed to break through the encirclement to the south. But the Allies had to leave their artillery, trucks and tanks behind. In total, the Germans took 1,200,000 men prisoner. Even later on the Eastern Front, never would so many men be marched off into captivity at once as after the British-French debacle of May 1940.

In the final phase of the campaign, German units turned south to neutralize what was left of the French armies. These were tense days but militarily of minor importance. For a month there

Captured French officers. On the right is Major General Veiel, commander of the 2nd Panzer Division.

was still bitter fighting in the rest of France, but the battle was essentially over. On 17 June the French offered peace terms, and they were spared no humiliation. The armistice was signed in the same railway carriage in

Guderian discusses the situation in his SdKfz 250. His adjutant keeps an eye on movements in the air, and for good reason: due to the rapidity of troop movements, the Luftwaffe often bombed its own men.

which the Germans had had to accept defeat twenty-two years earlier. The Wehrmacht had achieved its greatest victory at an astonishingly low cost. The Germans counted 49,000 dead and missing, while the French suffered 120,000 casualties, figures that give a good idea of the ferocity of the fighting during these six weeks in May.

During this period, Guderian also had the opportunity to examine the captured equipment more closely. When he discovered that the French Char Bs could easily withstand direct hits from the German 37mm anti-tank guns, he realized that if these tanks had been used correctly they could have cut through the corridor without any problems. The tests also showed that not only were the German tanks under-armoured and under-gunned compared to the Char B, but also that the German anti-tank guns had to be completely overhauled. But this observation was only a foretaste of what was to come: in 1941, the German Army in the Soviet Union would be confronted with tanks that were even better armoured and armed.

Chapter 9

The Eastern Front

Operation Barbarossa

After the successes of 1940, Hitler turned his attention to the Soviet Union, which he saw as his major political and military/economic opponent. However optimistic Hitler might have been after the western campaign, the scale of the task that awaited the Wehrmacht could not be underestimated. Most important of all, the population of the Soviet Union was much larger than that of Germany: even taking into account the unreliability of Soviet statistics, the USSR in 1941 had at least 147,000,000 people, while in 1939 Greater Germany had only 83,670,000. The Germans would have to deploy all their available men in an attack on the Soviet Union, while the Soviets would still have millions of reserves at their disposal. It was clear that the Wehrmacht should not get caught up in a war of annihilation, because they simply did not have the men for it. It was also necessary to avoid being swallowed up by the huge distances of the Soviet Union. If the Red Army were able to withdraw to the East in an orderly manner, insurmountable problems would arise. If, on the other hand, the cohesion of Soviet forces could be broken, the army would disintegrate and a German advance would become unstoppable. The bottom line was that victory had to be achieved in the first weeks of the campaign, just as in France. Operation Barbarossa was built on this assumption.

A massive attack aimed at Moscow, combined with a northern and southern flanking movement, would have to break the Red Army before the Dnieper-Dvina line, within 500km of the Polish-German border. This distance of 500km was crucial, because beyond this point the logistics would become unmanageable. Such limitations on the new mobile style of warfare had played a much less important role in the 1940 campaign in the West, because the depth of the attack was never more than a few hundred kilometres. The entire campaign in the West could have been supplied by a chain of trucks travelling back and forth from the German border. Based on the experience in France, the Wehrmacht's logistical staff had calculated that the most efficient range for supply trucks was 600km, in other words an operational depth of 300km. Beyond this distance, trucks themselves used so much of the fuel they transported that this mode of transport quickly became inefficient. To accommodate this, the Wehrmacht split the truck fleet into two: one part travelled with the armoured units and supplied them with fuel and ammunition from intermediate depots. Another part drove to and fro from the Polish-German border to maintain supplies in these depots. With these logistical adjustments it was hoped that the operational depth could be extended to 500km, precisely the Dnieper-Dvina line. This depth was mandatory: any further and the Wehrmacht's logistical problems would multiply exponentially.

German preparations

Hitler laid down the principles for Barbarossa in Directive 21 of 18 December 1940. However, this Directive did not specify a target, such as a city or a region that had to be captured. In Hitler's mind, there would simply be a massive confrontation between the Germans and Russians at the border, resulting in the destruction of the Red Army. Armoured columns would then be able to penetrate deep into the Soviet rear and make it impossible for any Russians who had survived the initial fighting to withdraw. The final line that the German troops had to reach ran from the Volga to Archangel. The ultimate goal of the operation was the subject of heated discussions between Hitler and the OKH in 1940 and early 1941. The targets were, successively, Moscow, Leningrad and Kiev. On 26 July 1940, the OKH had proposed Moscow as the target, but subsequently, Hitler would often suggest Kiev and/or Leningrad. All three cities were important to the Soviet Union. Kiev, the capital of Ukraine, controlled rich grazing and arable areas, Moscow was the administrative and logistical centre, and Leningrad, the former capital, was a warm-water port.

This controversy over its ultimate goal cast a shadow over the operation. Hitler and the OKH were thinking along different lines, and the OKH did not fully support Hitler's plans. Moreover, the lack of a clear goal meant that the OKH had to divide its forces. The panzer divisions were central to the plan, but given its ambition, the number of panzer divisions available was insufficient, despite the fact that their number had almost doubled since the campaign in Western Europe. However, there was a catch: the original number of ten panzer divisions in 1940 had apparently doubled by the start of Barbarossa, but this was achieved by halving the number of tanks per division. While the original panzer divisions had two tank regiments with a total of 280 tanks, the panzer divisions of 1941 had only one regiment with 140 tanks. Each regiment had two units; only three of the twenty new divisions had three units. On balance, the number of tanks had not increased compared to May 1940, but the tank strength was distributed over more units. This weakening was partly offset by replacing the PzKpfw I and II tanks with PzKpfw III and IV. Czech PzKpfw 35(t) s and PzKpfw 38(t) s were also available, which had proved to function excellently in practice. In addition, captured French materiel was distributed among the newly established divisions, but this proved completely unsuitable for the terrain and climate of the Eastern Front, and its use would contribute greatly to the confusing situation over spare parts.

In addition, the Wehrmacht had taken scarcely any time between the campaign in the West and the launch of Barbarossa to evaluate the former campaign and to raise the professionalism of command at the tactical and operational level. Officers did receive extensive additional training at special cadre schools, but they were unable to train sufficiently with the divisions they were part of. Moreover, men rotated between military service and civilian life during the year, and it had not been possible to build up a sufficiently large body of trained reservists. The preparations were far from perfect in comparison with the campaign in the West, and Barbarossa was late in getting started. The aim was to get underway as soon as possible after spring and the accompanying mud, and 15 May was generally seen as an ideal date. But military actions in the Balkans forced Barbarossa to be postponed until 22 June.

Composition of the German armed forces

The German forces were concentrated in three Army Groups:

- Heeresgruppe Nord
- Heeresgruppe Mitte
- Heeresgruppe Süd

The Pripet Marshes formed a natural barrier between Heeresgruppen Nord and Mitte on the one hand and Heeresgruppe Süd on the other.

The composition of the force was as follows:

Heeresgruppe Nord under von Leeb consisted of seven infantry divisions and Panzergruppe IV with three panzer divisions under Hoepner.

Heeresgruppe Mitte under von Bock consisted of forty-two infantry divisions and nine panzer divisions, the latter divided as follows:

- Panzergruppe II under Guderian with five panzer divisions
- Panzergruppe III under Hoth with four panzer divisions.

Heeresgruppe Süd under von Rundstedt consisted of fifty-two infantry divisions and five armoured divisions, including:

- Panzergruppe I under von Kleist with five armoured divisions
- The 3rd and 4th Romanian Armies (fifteen divisions in total)
- A Hungarian motorized corps consisting of two divisions
- An Italian corps

The allocation of the large number of armoured divisions to Heeresgruppe Mitte shows the OKH's preference for Moscow as the final objective of the offensive.

The Wehrmacht in 1941

Heeresgruppe	Composition
Heeresgruppe Nord under von Leeb	7 infantry divisions Panzergruppe IV (Hoepner), 3 panzer divisions
Heeresgruppe Mitte under von Bock	42 infantry divisions Panzergruppe II (Guderian), 5 panzer divisions Panzergruppe III (Hoth), 4 panzer divisions
Heeresgruppe Süd under von Rundstedt	52 Infantry divisions: • 35 German infantry divisions • The 3rd and 4th Romanian armies (15 divisions in total) • A Hungarian motorized corps consisting of 2 divisions • Italian Corps Panzergruppe I (von Kleist), 5 panzer divisions

At the beginning of Barbarossa, the Germans had a total of seventeen panzer divisions, with two in reserve. These divisions had the following tanks at their disposal:

- 410 PzKpfw Is
- 746 PzKpfw IIs
- 149 PzKpfw 35(t) s
- 623 PzKpfw 38(t) s
- 965 PzKpfw IIIs
- 439 PzKpfw IVs.

In total, this amounted to 3,332 tanks. Many of these were, however, light and outdated models compared to the Soviet tanks they were facing. The rest of the Germans' tanks were in Western Europe or North Africa. On paper, the German Army looked impressive. It consisted of well-trained soldiers led by a capable corps of officers. Their equipment was in many ways superior to that of their opponents and they had a tried and tested doctrine that had proved its worth in previous campaigns. The officers and men had shown during the campaigns in Poland and Western Europe what a well-trained and well-led army could do. However, the majority of the German force was not much more mobile than the Soviets. Most of the artillery was pulled by horses, not trucks or half-tracks. Of all 139 divisions deployed during Barbarossa, only the seventeen panzer divisions and the ten motorized infantry ones (including two SS divisions), some 18 per cent of the force, could actually be called motorized; the rest of the units had to move on foot or by horse and cart.

Composition of the German Panzerwaffe

However impressive the performance of the Panzerwaffe (tank force) had been in the past years, the fact remained that many of the tank designs in 1941 were outdated and not up to the task that lay ahead. The lion's share of the 4,700 tanks in total that the Germans had at their disposal were PzKpfw IIIs, supplemented by the newer PzKpfw IV, as well as the outdated types I and II and the Czech tanks 35(t) and 38(t). The composition of the tank force deployed for Barbarossa did not differ significantly in terms of structure from that of the total force, as shown below.

Composition of the Panzerwaffe in 1941

Type	Numbers	Percentage
PzKpfw I, II, 35(t) and 38(t)	2,710	57.6
PzKpfw III	1440	30.6
PzKpfw IV	550	11.8
Total	4,700	100.0

In 1940, the Germans realized that the French had better tanks than they did and that it was due to the incorrect use of these, rather than the performance of their own tanks, that the Panzerwaffe was so successful. This would happen again in the Soviet Union. The initial successes of the German armoured units masked the fact that a large proportion of their tanks were simply no match for Soviet tanks such as the T-34 and KV-1, as we shall see.

Within the German tank force, the role of the Czech PzKpfw 38(t) tank, produced by the Skoda works, is often neglected. This tank accounted for between 20 and 24 per cent of the Panzerwaffe in the period 1939 to 1941, a period in which the PzKpfw IV, for example, played only a limited role.

This would change in the years to come, when the PzKpfw IV would become the workhorse of German armoured units. The PzKpfw 38(t) was not taken out of production until 1942, and its chassis remained in use as a platform for anti-tank guns in the form of the Marder III and the Hetzer.

Annual German tank production during Barbarossa was around 1,000 units. This number pales into insignificance in comparison to the production figures of the Soviet Union, where, for example, an astonishing 24,000 tanks were produced in 1941 alone. Guderian noted in his memoirs that in 1933 he had visited a Soviet tank factory that produced twenty-two tanks of the Christie-Russky type per *day*. In his book *Achtung-Panzer!* he estimated annual Soviet production at 12,000, much to the chagrin of the high command. It was clear that the Soviet Union was a formidable opponent when it came to tank numbers. Only after the start of Barbarossa did it become clear that the quality of some Soviet tank designs also exceeded that of the Germans'.

German tank production 1938–1941 by type

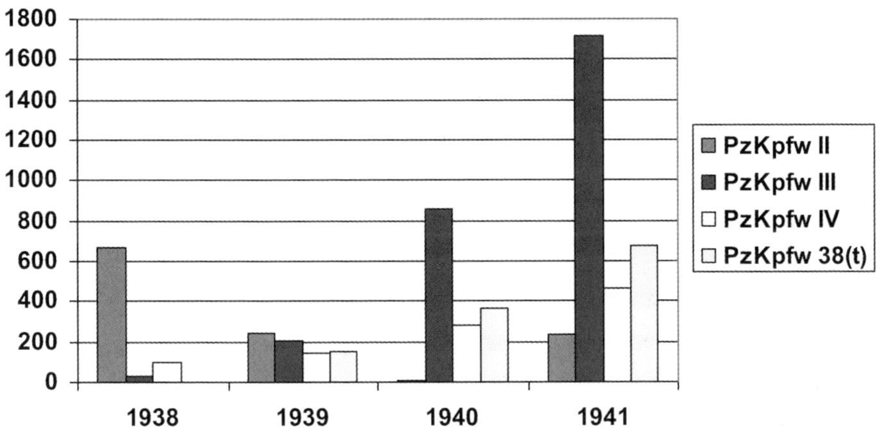

The above clearly shows the emphasis on production of the PzKpfw III and the PzKpfw 38(t) in the period up to 1942. It was not until 1941 that substantial growth was seen in the production of the PzKpfw IV, the only tank that proved more or less equal to its Soviet counterparts. However, it was not until 1942 that this tank would be able to use the 75mm/L43 or L48 gun, the armour-piercing shells of which could knock out heavily armoured Soviet tanks such as the T-34. Until then, the task of halting these tanks rested on the shoulders of the PzKpfw III and the 88mm Flak gun.

It was not only in quality that the German tanks were inferior: by the second half of 1941 the army leadership realized that the losses of tanks were much greater than initially expected and that they were faced with a longer campaign than anticipated. And even more importantly, they realized that further improval of the current models was not the solution; a new generation of tanks simply had to be developed. The Soviet attitude towards German tanks can best be illustrated by the visit of a Soviet delegation to German factories and training institutions in the spring of 1941, not long before the launch of Barbarossa. Hitler had explicitly stated that the visitors should have access to all facilities and that nothing should be hidden from them. However, the Soviet officers refused to believe that the PzKpfw IV was the heaviest tank in the German arsenal and complained that Hitler's orders were not being carried out. They made this point so often and so emphatically that the Germans became extremely worried: everything indicated that the Soviets were in all probability already developing a new generation of tanks. The appearance of the T-34 at the front in the summer of that same year confirmed their worst fears. And as we shall see, the Soviets had more unpleasant surprises in store in this field.

Soviet tank doctrine

As we have seen, in the early 1930s Marshal Tukhachevski had been the great architect of Soviet armour, creating formations that bore a strong resemblance to the German armoured units. After he fell victim to the great purges of 1937/1938 he was succeeded by Marshal Kulik, who had his own ideas about the use of tanks and ordered a temporary halt to the production of anti-aircraft and anti-tank guns. The position of the tank units was further undermined by the views of General Pavlov, the commander of the Soviet tank units deployed in the Spanish Civil War. He saw no role for large, independently operating tank formations; in his eyes they were only there to support the infantry. For Kulik this was another argument to disband the existing tank units.

This development was comparable to the views aired in France at the same period, but during the Winter War against Finland and the campaign in Poland, this concept failed to prove its worth. On the other hand, at the front in the east of the Soviet Union against Japan, Zhukov achieved great success by the concentrated use of independent tank units. Lessons from the latter campaign, and from the German invasion of Poland and Western Europe, led Stalin in 1940 to reinstate the mechanized corps. These were composed of:

- Two armoured divisions composed of two tank regiments of 200 tanks each, a regiment of motorized infantry and an artillery regiment
- A motorized infantry division. The armoured divisions had a wide variety of more or less useful tanks at their disposal, increasingly also T-34s and KV-1s.

The major problem of these large and cumbersome tank corps was direction and coordination, especially because they generally carried no radio equipment for mutual communication. The Soviets could not in any way match the C3I capacity and manner of operating in fluid, mobile operations of their German opponents; their operations most resembled that of tank units in the First World War. Only gradually as the war progressed would the use of radio increase.

The Russians solved their communication problems by practising prescribed manoeuvres. Soviet tank attacks always took place frontally and had limited objectives, so they were unable to respond to new situations or form temporary battle groups in the style of the German Kampfgruppen. If the objectives were not reached the first time, a second attack along the same lines generally followed the next day. This continued until the objectives were achieved or all men and materiel were simply expended. In addition, the Soviet tank corps had no efficient system of fuel and ammunition supply, and recovery and repair of tanks in field workshops was unknown, as was any system for evacuating the wounded: all necessary preconditions for the deployability of tank units in the field.

All this meant that the Soviet tank units would play no significant role in the opening rounds of the war. However, they would rise impressively from the ashes in late 1942 and play a central role in the counter-attack and advance into Germany.

Soviet preparations for war

The Soviets were by no means unprepared for an attack by the Germans. They had long anticipated it and had built up a large military force to see them through the first year of such a war. It was also no coincidence that at the start of Barbarossa the bulk of the enormous Red Army was located within 60km of the German border. The Soviet economy had in fact been on a war footing from the first Five-Year Plan in 1928, with a major increase in arms production in the third Five-Year Plan from 1938.

Forecasts of the Five-Year Plan 1938–1943

Tank types	1938	1943
Light tanks	16,000	24,000
Medium tanks (T-26, T-46)	14,000	21,000
Fast tanks (BT-7, BT-8)	5,000	15,000
Heavy tanks (T-28, T-29, T-35)	400	775
Total	35,400	60,775
Aeroplanes	20,500	50,000

After 1939, military expenditure rose dramatically, from 25.6 per cent of GNP in 1939 to 32.6 per cent in 1940 and 43.4 per cent in 1942. As a result, the already impressive Soviet arsenal in 1939 was supplemented by no fewer than 17,000 aircraft, 7,600 tanks, 80,000 pieces of artillery and mortars, and more than 200,000 machine guns and automatic weapons.

Growth of the Red Army 1939–1941 compared to the German Army

	Red Army 1 January 1939	Red Army 22 June 1941	German Army 22 June 1941
Divisions	136	303	155
Men	1,943,000	5,710,000	4,600,000
Artillery, mortars	55,800	115,900	43,407
Tanks	18,400	23,106	3,998
Aeroplanes	17,500	22,400	3,904

While about 20 per cent of Soviet equipment was obsolete, 80 per cent was modern and included designs that were superior to anything the Germans could field.

Soviet tanks

We have seen that the Soviet Union had devoted much attention to the mechanization of its armed forces in the 1930s. However, many of the Soviet tanks from before 1939 were based on hopelessly outdated concepts. Tanks such as the SMK, the T-100 and the T-35 had to fight their way across the battlefield like lumbering land battleships, comparable to the French Char Bs. There were also better tanks such as the T-28 and the T-26, the latter already labelled as effective by the Germans in Spain, and the BT series, fast, lightly armoured but well-armed. Then there were amphibious tanks, the T-37, T-38 and T-40.

The T-26
The Soviets designed the T-26 as a light infantry tank to perform reconnaissance missions and operate in the vanguard. It was virtually a copy of the British Vickers 6-ton tank, but at 30kph it was not really fast enough for reconnaissance missions. The T-26 had thin armour, a 45mm gun and two machine guns. With its three-man crew, the tank and its armament posed a threat to the PzKpfw III and IV, but its thin armour made it very vulnerable. The T-26 was first used on a large scale in the war against Finland and later on the Eastern Front.

Among these more or less useful and sometimes bizarre designs, however, were tanks that would prove their worth in the coming period: the T-34 and the KV-1 (named after one of Lenin's loyal followers, Klimenti Voroshilov), designed by Koshkin and Kotin respectively. The T-34 featured a potent 76.2mm gun, a Christie spring system, and its front had such an angle that its 45mm armour matched that of a tank with 90mm hull armour. The KV-1 also had a 76.2mm gun, with 90mm armour and a traditional spring system. Both tanks had the same reliable aluminium diesel engine and wide tracks, which provided much better manoeuvrability in mud and snow than their German opposite numbers. Both also had enough space to accommodate a heavier gun in the future.

In the war with Finland, the Soviets gained valuable experience in using tanks, and as a result, the newly developed KV-1, KV-2 types, as well as the T-34, were put into production at an accelerated pace.

BT-7M

With its 500hp engine and top speed of 86kph, the BT-7M was the racehorse of the Soviet armoured divisions. The tank was armed with a powerful 76.2mm anti-tank gun and two 7.62mm machine guns. The three-man crew was protected by thin armour, the tank's weak point. Nevertheless, this type remained the backbone of the Soviet armoured units for a long time.

Its planned successor was the BT-IS, a tank that never got further than the drawing board. Instead, the engineers involved designed one of the most famous and effective tanks in history, the T-34.

At the start of Barbarossa, mass production of these tanks was well underway, and large numbers were in operational service with various units. Their presence was an unwelcome surprise to the Germans, as was the KV-2, an assault tank armed with a 152mm (!) howitzer. This generation of Soviet tanks far outshone existing German tanks in terms of armour and armament, and more worryingly, the Soviets were already experimenting with a new generation of tanks which would be even more heavily armoured and armed. This was based on the mistaken assumption that the Germans were increasing the armour of the existing PzKpfw III and IV designs from 50 to 100mm.[3] The consequences would soon become apparent: the Germans would be confronted in the coming period with an opponent of a completely different calibre to any they had faced up to that point. They were well aware that, to win the war, eliminating Soviet production capacity was essential, no matter how outdated the tank designs were.

---

3. This unfounded assumption was made by Marshal G.I. Kulik, a protégé of Stalin. The point was that all tank design and production efforts were aimed at surpassing the Germans. Kulik was later dismissed by Zhukov as incompetent.

Operation Barbarossa

At 3.15am on 22 June 1941, Operation Barbarossa broke with full force on the Soviet Union. Despite all the signs that preceded it, the Soviets were taken completely by surprise, and the Germans broke through their lines with little trouble. The Soviet command structure collapsed almost completely, and any form of direction and coordination soon disappeared.

OKW and OKH

At the top of the German Army, two institutions were of crucial importance: the *Oberkommando des Heeres* (OKH) and the *Oberkommando der Wehrmacht* (OKW).

In the OKH, General von Brauchitsch was the supreme commander (*Oberbefehlshaber*) and General Halder was the chief of staff.

Von Brauchitsch advised Hitler and was responsible for all parts of the army, especially for personnel, the Feldheer (field army) and the Ersatzheer (reserve).

As chief of staff, Halder led the army in the field and was responsible for combat operations, gathering intelligence, troop movements, etc. In addition, the General Staff and all kinds of other more specialist departments came under his control.

A competing body was the OKW, which covered not only the army, tbut also the air force and the navy. It was headed by Hitler and had Keitel as chairman (also called 'Lakeitel' [lackey Keitel] by some to indicate his servile relationship to Hitler). The OKW rarely met with Hitler present (only three times up to the end of 1941), since Hitler preferred bilateral consultations.

The influence of the OKW grew over time. In the period from September 1939 to early 1941, the OKH planned the campaigns with the help of the other armed forces. This changed after 1939. The OKH planned the war in Western Europe, while the OKW was given responsibility for the campaigns against Norway and Denmark. In the spring of 1941, when the OKH was planning Barbarossa, the OKW was given responsibility for all 'quiet' areas, including North Africa. In the period after that, the

OKH focused on the war in the East, while the OKW, with Chief of Staff General Jodl, was responsible for all other theatres. The influence of the army decreased as a result, and after the forced departure of von Brauchitsch at the end of 1941, more and more areas which had been the responsibility of the OKH, such as personnel, were transferred to the OKW. Towards the end of the war, this also applied to the General Staff. Despite the existence of these parallel structures, there were few major problems in practice, because the staff officers, with their shared background, ensured that activities were properly coordinated.

Barbarossa's goal was to push back the Red Army behind the Volga along the Archangel-Astrakhan line. Above all, the Russians had to be prevented from retreating into their Asian hinterland, as they had done successfully in 1812. The Red Army had to be neutralized in a number of encirclement actions, in which the armoured troops would play a crucial role. In what

Guderian at Brest-Litovsk at the beginning of Operation Barbarossa. On the right is his chief of staff, Colonel Freiherr von Liebenstein.

follows, we will discuss the activities of the various Army Groups and Guderian's Panzergruppe II.

Heeresgruppe Nord

Heeresgruppe Nord had Panzergruppe IV under Hoepner at its disposal. This Panzergruppe in turn consisted of two Panzer Corps, LXI Corps under Reinhardt and LVI Corps under von Manstein. Apart from the panzer divisions, these were composed of motorized and non-motorized infantry. The two corps were followed by the SS Division Totenkopf, which was to follow whichever corps made the most progress.

The main objective of Heeresgruppe Nord was Leningrad. Speed was of the utmost importance, and Leningrad had to be captured as quickly as possible so that the units involved could swing south to support the offensive of Heeresgruppe Mitte towards Moscow. A first major obstacle was the crossing of the Daugava River, some 300km east of the border. If the Soviets were able to blow up the bridges, the offensive in the north would be seriously delayed, with all the consequences that entailed. One of the first surprises the Germans faced were the Soviet heavy tanks of the KV-1 and KV-2 types. A single tank was sometimes able to hold up an entire division, and they could only be taken out by deploying the famous 88mm Flak gun. Despite the delays this caused, the Germans were able to

Armoured units with the characteristic 'G' advancing ever deeper into the Soviet Union. Guderian, like all German commanders, was often to be found in the vanguard, constantly keeping abreast of events on the ground.

capture the bridges over the Daugava at Daugavpils intact on 26 June, only four days after the start of Barbarossa.

The Panzer Corps of Reinhardt on the left wing and that of von Manstein on the right then advanced rapidly with the aim of capturing the bridges over the Velikaya. Between them and this river was the Stalin Line, fortifications west of the Velikaya, which they broke through successfully. Then, on 4 July, Reinhardt succeeded in capturing the bridges over the river at Ostrov, and von Manstein captured the bridges at Opochka on 6 July. The capture of the Stalin Line also gave them access to the two important swamp crossings that shielded the south of Leningrad. Only the Luga River now lay between the Germans and Leningrad itself.

The German advance stalled, however, at this point. Reinhardt's units on the left wing had great difficulty with the terrain, and von Manstein discovered that the area between the Luga and Lake Ilmen had been heavily fortified by the Soviets. A plan was drawn up for von Manstein to launch the main attack in the south by crossing the Luga, supported by an attack by Reinhardt in the north-west. On 14 July, Reinhardt's units succeeded in crossing the Luga and then the Zabsk further east, which put them in an excellent position to launch an assault on Leningrad itself. In the meantime, however, von Manstein had not yet succeeded in crossing the Luga. A kind of stalemate now arose: Reinhardt, instead of advancing further, waited for von Manstein. If von Manstein had successfully launched an attack, then some of the units still opposite Reinhardt would undoubtedly have been redeployed to counter von Manstein, and this would have improved Reinhardt's chances of advancing directly on Leningrad. But now the Soviet defenders opposite Reinhardt had plenty of time to dig in and bring up reinforcements. When Reinhardt was given the green light, almost three weeks later on 8 August, he was confronted with stubborn resistance. Von Manstein's units were repositioned further west, but although an attack on 10 August led to nothing, Reinhardt had meanwhile succeeded in breaking through the Soviet defences and occupying the area between the marshes and the city on 14 August. Von Manstein's units were now moved west to support Reinhardt, but on 19 August they clashed with the Soviet Thirty-Fourth Army, which threatened the German flanks. By 23 August von Manstein had managed to neutralize the Thirty-Fourth Army and was able to advance further. In the meantime, his infantry had taken Novgorod and was able to advance unhindered towards Leningrad.

All in all, however, a month had been lost and the Soviets had been given a chance to regroup. It dawned on them that it might be possible to defend the city against the Germans, and their morale rose visibly. The delay seemed to have been fatal to the Germans, but by early September all their units were in their starting positions and the attack on Leningrad could begin. On 8 September, the Germans took Schlüsselberg, 40km east of Leningrad, and Leningrad was now cut off from the east. Only the resistance of the thirty-one encircled Soviet divisions could turn the tide. At this moment Hitler made one of his most controversial decisions: the city would not be taken by storm, but would be cut off and besieged. In his opinion, resistance would automatically collapse and Leningrad would fall into the hands of the Germans like a ripe plum. An important reason for this decision was that it was assumed that tanks would have to play an important role in the capture of the city by storm, whereas these tanks were urgently needed elsewhere, in accordance with the earlier plan, namely in the area of Heeresgruppe Mitte. In the meantime, the inhabitants of Leningrad and the surrounded troops refused to be intimidated: Leningrad would be besieged for a total of 900 days, but the city did not fall. It could only be supplied during the winter months over the frozen Lake Ladoga, but that did not prevent civilians and Red Army soldiers from successfully resisting the German attacks.

Heeresgruppe Mitte

Heeresgruppe Mitte benefited from the presence of two eminent tank commanders, Guderian and Hoth. In addition, it had the majority of the tanks, although, to a certain extent, it lacked an explicit goal: Moscow would only become the final target once Leningrad had been captured and a large part of Heeresgruppe Nord would then be available to reinforce Heeresgruppe Mitte. However, it was hoped that in the area of operations of Heeresgruppe Mitte large sections of the Red Army could be encircled and destroyed by Blitzkrieg-like tactics, thus delivering a decisive blow to the Soviets. This plan worked very well, although the infantry could not keep up with the rapidly advancing armoured units and slowed down the advance.

Guderian himself was not informed of Hitler's plans to invade the Soviet Union until November 1940. He had some reservations about conducting a war on two fronts, but his objections, like those of many others within the Wehrmacht, fell on deaf ears. Despite his misgivings, Guderian took a full part in the discussions about the strategy to be pursued in early 1941. He was convinced that the only correct plan was to capture Moscow, the capital of the Soviet Union. In his opinion, this would cause the communist regime to collapse and the war to be over quickly. For Barbarossa, Guderian was given command of Panzergruppe II, which consisted of three motorized corps with a total of five armoured, one cavalry and three motorized divisions. Together with Panzergruppe III under Hoth, Panzergruppe II was stationed in Poland as part of Heeresgruppe Mitte under von Bock in preparation for the offensive. The mission of both Panzergruppen was to break through the Soviet lines, destroy the enemy in a number of encirclement battles and advance deep into Soviet territory towards Smolensk. In total, Guderian had fewer tanks at his disposal for this mission than he had had a year earlier during the campaign in the West.

Before Barbarossa was launched, Hitler summoned all the major commanders on 14 June to explain his reasons for attacking the Soviet Union and to discuss the latest reports on the preparations for it. There was no room for discussion, and the assembled group of officers could do nothing but listen to the Führer. Guderian returned to Poland in a gloomy mood; a war on two fronts seemed impossible to win, given the circumstances. At that time, the Wehrmacht was divided between Europe and North Africa, which meant that only 145 of its 205 divisions were available for the camapaign against the Soviet Union. In addition, Guderian had already pointed out the enormous manufacturing capacity of the Soviet Union. The reports of the German military attaché in Moscow, General Köstring, had also painted a threatening picture of the military strength of this enormous country. Hitler had paid little attention to these reports, or to accounts of the production capacity of Soviet industry and the stability of the communist regime.

Guderian's starting point was Brest-Litovsk on the Bug. He had taken the fortress of Brest-Litovsk during the Polish campaign, which is why he assumed that he knew exactly how to tackle it again. He proposed that his panzer divisions cross the Bug on both sides of the fortress, and the infantry tackle the fortress itself. The Fourth Army under von Kluge was to provide the necessary infantry, which was to follow Guderian's Panzergruppe during the advance and provide infantry and artillery support during the crossing of the Bug.

In order to guarantee unity of command, Guderian was prepared to place himself under the command of Field Marshal von Kluge. During the preparations for the campaign, however, Guderian immediately found himself in conflict with von Kluge about the strategy to be followed. As we have seen, Guderian and von Kluge had already had a major disagreement in Poland. Von Kluge wanted the infantry units to first force a breakthrough, before the armoured units were deployed. The armour would then advance through the gap that had been created to objectives located deeper behind the front. Guderian, however, wanted to use the element of surprise to the fullest extent possible by deploying his armoured units immediately, in order to force a breakthrough as quickly as possible, and use their speed to advance deep into the hinterland. He assumed that capturing the fortress of Brest-Litovsk would take some time. Guderian immediately presented this difference of opinion to von Bock, commander-in-chief of Heeresgruppe Mitte, who gave him free rein to deploy his units as he saw fit, which he did.

The conflict with von Kluge, which would continue to haunt Guderian for a long time, was based on both personal and professional resentment. Von Kluge saw Guderian as a dangerous, impulsive innovator; Guderian, in turn, saw von Kluge as an inflexible traditionalist, an exponent of the old school of infantry tactics and an opponent of mobile warfare. Guderian had disliked von Kluge from the first and became increasingly suspicious of his intentions. Eventually, the two officers would disagree so violently that von Kluge, despite Guderian's peace overtures, ultimately saw no other way out than to challenge Guderian to a duel.

A few days before the offensive, Guderian and his fellow officers were surprised by two orders from the OKW. The first indicated that in the event of violence against the civilian population or prisoners of war, the soldiers responsible should not be automatically prosecuted under military law. Instead, punishments should be imposed by the officer in command of the men involved. This order, in Guderian's eyes, would place officers in an impossible position in relation to their subordinates, and vice versa.

Von Brauchitsch, as commander-in-chief of the army, had immediately recognized the dangers of this procedure and had added that this order should only be executed if discipline would not be jeopardized. Guderian refused to distribute the order to the commanders of the various divisions under his command and passed this refusal on to von Bock. Von Bock, in turn, had in the meantime refused to pass on a second order, the so-called *Kommissarbefehl*, to the commanders of Heeresgruppe Mitte. This second order concerned the treatment of political commissars who had been taken prisoner: they could be executed without trial. In his memoirs, Guderian regretted that neither the OKH nor the OKW, as the primary responsible parties, had prevented these orders from being drawn up and disseminated. Guderian argued that the fact that the Soviet Union had ratified neither the Hague Convention nor the Geneva Conventions, and it was therefore completely unclear which laws of war they would adhere to, was not in itself a reason to ignore the obligations arising from these treaties vis-à-vis the Soviets. As it turned out, in the harsh reality of the Eastern Front, both armies would soon show little concern at all for any laws.

Soon after hostilities began in the early morning of 22 June, Guderian's first units crossed the Bug. Guderian himself followed at 6.50am, and

Operating up front was not without danger. Here a bomb has hit the road a few hundred metres ahead of Guderian.

his *Gefechtsstaffel* (staff unit) consisting of two communication vehicles, reconnaissance vehicles and motorcycles followed at 8.30am. Together they followed the tracks of the tanks of the 18th Division that had advanced rapidly

A day in the life of Heinz Guderian

On 24 June he left his headquarters at 8.25am and headed for Slonim, where the 17th Panzer Division had arrived. Between Rozama and Slonim his staff unit was fired upon by Soviet infantry. A battery of the 17th Panzer Division and motorcyclists returned fire, but without much effect. Guderian personally fired at the enemy with the machine gun on his half-track vehicle. After some time he succeeded in driving off the enemy units and was able to report to the headquarters of the 17th Panzer Division in Slonim at 11.30.

During the meeting with General von Arnim, the divisional commander, Guderian suddenly heard rifle and machine-gun fire. A burning truck blocked the view, so it was unclear what was happening, but then two Soviet tanks suddenly appeared out of the smoke. They tried to break through to Slonim, firing continuously and being pursued by German PzKpfw IVs. When the tanks saw the officers, they immediately started shooting at them at very close range. Everyone dived to the ground, except for one officer who was wounded. The tanks managed to break through but were eventually destroyed in the centre of the city.

Guderian then visited the front at Slonim and drove a tank through no-man's-land to the 18th Panzer Division. At 3.30pm he was back in Slonim after giving the 18th Panzer Division and the attached infantry division the necessary orders.

The journey back to headquarters continued, but not without problems: at the town of Slonim, Guderian came across Soviet infantry climbing out of trucks. He ordered his driver to drive at full speed through these troops, who were so surprised that they had no time to grab their weapons. He was recognized, however, and his death was reported in the Soviet press a few days later.

At 8.15pm Guderian was back at his headquarters to evaluate the battle situation and make plans for the coming day.

eastward. Guderian accompanied the forward units of this division together with General Nehring until late in the afternoon. The following five days passed at breathtaking pace, described by Guderian with great verve in his memoirs. Despite the enormous width of the front – often many hundreds of kilometres – Guderian was always '*an der Spitze*' (at the head), leading his units at the tactical and sometimes operational level. He solved the problem of the great distances involved simply by demanding twice as much effort from himself, his drivers and his staff. He came under enemy fire regularly and on one occasion barely escaped with his life. But luck was always on his side and he was not wounded even once.

In the days that followed, Guderian's units advanced rapidly toward Minsk. On 27 June the first tanks of the 17th Division reached the outskirts of the city and made contact with the tanks of Hoth's Panzergruppe III, which had taken a more northerly route. The spearheads of Heeresgruppe Mitte had covered 360km in five days, and the two arms of the pincer movement had encircled countless Soviet formations. These deep strategic penetrations had left behind large numbers of enemy, whose mere presence had tied up substantial German units. The German infantry proved unable to keep up with the pace of the advance while at the same time encircling and capturing these enemy troops.

The following week, Guderian's units had two main objectives. First, he and Hoth were to actively reconnoitre the route to Smolensk and the Dnieper. Second, his units were to prevent the encircled Soviets from breaking out. As Guderian pushed further and further east, the situation behind the front became increasingly opaque. Guderian believed in safety through movement, a concept that the slowly advancing infantry found increasingly difficult to appreciate, as they were constantly confronted by Soviet units appearing here and there. Von Kluge demanded in increasingly forceful terms that rest periods be arranged to re-establish contact with the infantry and to allow them to clear the various pockets of fierce resistance, some of which were still deep in the west of the Soviet Union. Guderian, however, continued to look to the east.

When von Kluge's pleas seemed to have fallen on deaf ears, he sternly ordered Guderian to call a temporary halt in late June. Guderian immediately contacted Hoth, to see how they could evade this order and advance together to Smolensk in defiance of von Kluge. Summer rains, meanwhile, came to the aid of the Soviet defenders, and as the rudimentary road system turned into a quagmire, the German advance slowly but surely lost momentum. Ultimately, in the following days, millions of Soviets were taken prisoner in four separate encirclement battles: at Brest-Litovsk, where the garrison of the fortress had stubbornly held out for several days; at Bialystok, where six Soviet divisions were encircled; near Volkovysk, where a furher six were surrounded; and, the largest encirclement of all, at Minsk, where no fewer than fifteen divisions were taken. But unlike the French in 1940, the Soviets had no intention of surrendering and continued to fight bravely. At times they managed to break through the thin inner ring of German infantry, only to be driven back by the tanks of the outer ring. Hundreds of thousands of men, however, managed to evade capture and fought their way back through the lines, or were completely absorbed into local communities, where they would form the basis of future resistance groups. This process would be accelerated in the coming period by the murderous activities of the various Einsatzgruppen, the secret police and the Sicherungs Divisions that followed the advancing Wehrmacht units into the conquered area.

Guderian, meanwhile, was annoyed by time that these clearing operations took; he wanted to free his tanks from these static operations as quickly as possible and have them advance some 350km further along the Smolensk – Elnya – Roslavl line towards the crossings of the Dnieper. On 1 July, von Kluge discovered that, despite his orders, elements of Hoth and Guderian's 17th Panzer Division were still on their way to the Berezina, one of the larger rivers west of the Dnieper. Von Kluge immediately called Guderian to order and was forced to do so again the next day, because the 17th Division had simply continued its advance. Guderian insisted to von Kluge that this was a case of communications breakdown, and the fact this had applied also to Hoth was just a coincidence. Von Kluge was not amused and threatened again to have both men court-martialled. That same day, Guderian's units also came up against T-34s for the first time. These tanks, with their low silhouette, heavy armour and formidable 76mm gun, would become a nightmare for the Germans in the coming period.

On the way to the Dnieper

After the last pockets of resistance at Bialystok had been cleared on 3 July, the first elements of the 3rd, 4th and 18th Panzer Divisions were able to move on to the Berezina, which they successfully crossed a short time later. The advance then continued at a rapid pace to the Dnieper, which was reached on 5 July. The Soviets launched a fierce counter-attack that was successfully repelled. Guderian was now faced with a dilemma. On the one hand, it was clear that the Soviets were sending more and more units to the Dnieper front, which they intended to defend at all costs. Establishing a bridgehead on the other side of the river, followed by a rapid action on the eastern bank, would frustrate Soviet aims. On the other hand, Guderian's position was particularly vulnerable: the rapid advance had left his supporting, non-motorized infantry units far behind, and his extended flanks were very vulnerable to counter-attacks. The infantry would need at least another seven to ten days to catch up, and until then the flanks would have to be protected by a weak screen of motorized infantry and reconnaissance units. To complicate things, the supply system was also in disarray. Consolidating the German positions was therefore a logical alternative. In addition, everyone was slowly but surely realizing that conditions in the Soviet Union were not comparable to those in France the year before. The operational strength of Guderian's panzer units was rapidly declining. The great distances, the poor condition of the roads, the dust and the mud meant that fewer and fewer tanks were operational. The mobile workshops were only prepared for a short, Blitzkrieg-like campaign, and the stock of spare parts was being rapidly depleted. There were no local facilities to conduct major overhauls, and most of the stranded vehicles could not be transported back to Germany due to the lack of transport and the pockets Soviet of resistance in the rear. For this reason, disabled vehicles were increasingly cannibalized to maintain the operational strength of the combat units.

Moreover, one of the previously recognized weaknesses of panzer divisions was becoming increasingly apparent: they were primarily designed for offensive actions and were less well equipped for defence. In the coming period, they would often have to fend off persistent attacks, without any support, from both encircled Soviet units and others seeking to relieve them.

Many of the bridges had been blown up by the Soviets during their retreat. The German engineers performed miracles, such as here where they have built two emergency bridges. A Flak 88 gun, the only answer to the heavily armoured Soviet tanks, is being carefully guided over the bridge.

Guderian's Panzergruppe was operating on a front four times as wide as that of the Polish and western campaigns of 1940, and the units were in danger of simply being swallowed up by the vast expanse of the country.

As expected, Guderian preferred a rapid crossing of the Dnieper, but von Kluge was clear: the armoured units should wait until the infantry arrived. Guderian stubbornly maintained his position and used the argument that every day of delay would give the Soviets the opportunity to strengthen their positions. The question was whether the German infantry would then still be able to break through the Soviet defences. Guderian used the well-known argument that speed in itself is a weapon because it creates uncertainty and confusion, and an opponent who has been confused can be defeated even by a weaker force. However, von Kluge remained steadfast until Guderian made it clear to him that 'his preparations were too far advanced to be undone … that the troops assembled in their starting positions could remain there for only a limited time because Soviet aircraft would soon discover and attack them.' Von Kluge, faced with a fait accompli, finally gave in and with the words 'your operations hang by a thread as always' he gave his permission.

On 10, 11 and 12 July Guderian's divisions crossed the Dnieper to advance on Smolensk. Guderian himself crossed on 11 July near Tolshino, where his headquarters were located and where Napoleon had set up his headquarters in 1812. The advance to Smolensk took four days, and the city fell to the 29th Motorized Division on 16 July. Hoth's Panzergruppe III had meanwhile encountered unexpectedly heavy resistance and had not yet managed to reach the city. Guderian and his Panzergruppe II faced three tasks in the following days:

- to prevent the encircled Soviets from breaking out to the east or south
- to make contact as quickly as possible with Hoth's units, which were forging a path from the north in a south-easterly direction towards Smolensk
- to try to gain ground to the east, towards Roslavl, as that city was a communication hub in the region. Elnya, further to the east, could eventually become the springboard for the attack on Moscow.

Understandably, Guderian was focused on launching an offensive towards Roslavl and Elnya, but Hoth was not able to make contact with Guderian's units until 26 July. The next day, Guderian received permission from von Bock to launch an offensive towards Roslavl as the first phase of an attack on Moscow. Guderian's units were reinforced for this purpose with three additional infantry corps, VII, IX and XX Corps; consequently, Panzergruppe Guderian was renamed 'Panzerarmee Guderian'. These infantry corps received special attention from Guderian in the following days because they had so far only seen sporadic action; he explained his method of attack, the way in which panzer and infantry units should cooperate and how communication should be managed between the various commanders. The latter was particularly important because these units had never worked so closely with tanks before.

But dark clouds were gathering above the Germans. In the top echelons of the Wehrmacht, the first hesitation became apparent about the next steps to be taken. Logistical problems were becoming increasingly severe. For example, ammunition for Guderian's forward units had to be transported by truck over a distance of 500km from the nearest depots. These in turn were

Contrary to the image many have of the Wehrmacht, its units often depended on horse traction. Here a horse-drawn light field howitzer passes men of the same unit.

many kilometres away from the nearest accessible junctions of the Russian railway network, which was rapidly being converted from broad gauge to European standard gauge. Guderian, who assumed that the German offensive would lose momentum if the advance to Moscow was interrupted, feared the army leadership would be tempted to interrupt the endless advance through the Soviet Union in order to bring in supplies and reinforcements. And there were also increasing indications that Hitler was considering a break in operations. Schmidt, Hitler's adjutant, alluded to this when he came to present Guderian with the Oak Leaves to his Knight's Cross on 29 July.

While the fighting around Roslavl was in full swing on 4 August, Guderian heard from Hitler himself in a meeting at the headquarters of Heeresgruppe Mitte that he was considering shifting the focus of the offensive to Ukraine, where Heeresgruppe Süd under von Rundstedt had not advanced as quickly as intended. Hitler described to the assembled company his immediate objectives. The industrial area around Leningrad was the first, but he had not yet decided which would come second, Moscow or Ukraine. He seemed to lean toward the latter for a number of reasons: first, Heeresgruppe Süd seemed close to victory there; second, he was convinced that the raw materials and agricultural production of Ukraine were necessary to the continuation of the war; and finally, he felt it was necessary to capture Crimea because from there the Soviets could threaten the Romanian oil fields. He hoped that Moscow and Kharkov could be captured by the beginning of winter.

Guderian emphasized during this conference that the engines of the tanks had suffered greatly from dust; as a consequence they had to be replaced as soon as possible. He also indicated that it was necessary to replace numbers of lost tanks. After some discussion, Hitler promised to deliver a total of 300 tanks to the Eastern Front, a number that Guderian considered completely insufficient. Hitler indicated that any newly produced tanks would only be given to recently formed units. In the ensuing discussion Guderian indicated that the great numerical superiority of the Soviets could only be dealt with if the losses of tanks were quickly made up.

At this point Hitler referred to Guderian's book *Achtung-Panzer!* saying, 'If I had known that the figures for Soviet tank strength that you gave in your book corresponded to reality, I would not, I believe, have ever started this war!'

Guderian had estimated in his book that at the time (1937) the Russians had 10,000 tanks; he had had great difficulty in getting this figure published, since Beck, the Chief of Staff of the OKH, fundamentally disagreed with it. However, Guderian had access to secret reports in which a figure of 17,000 was mentioned, and the numbers given in *Achtung-Panzer!* were therefore a conservative estimate. In addition to a commitment to provide the necessary tanks, Guderian obtained the concession that the bridgehead on the Desna at Elnya would be retained as a base for a future attack on Moscow. He decided to strengthen his position there by regrouping his units tactically and to wait for events to unfold. The fighting around Roslavl was decided in the Germans' favour around 8 August, after a total of 38,000 Soviet troops were captured, alongside 200 tanks and 200 artillery pieces.

Heeresgruppe Süd

Heeresgruppe Süd's advance had not got off to a smooth start. The left wing had been held up by the fortifications of Lvov and Prsemysl, and as a result, the right wing, which had encountered much less resistance, could not advance without exposing its own flank to counter-attacks. After the start of Barbarossa, the units on the left had only managed to advance 90km, while the other two Army Groups had already made much greater progress. The Stalin Line, which had caused little delay elsewhere, made the advance of Heeresgruppe Süd considerably more difficult. It was not until early July that the front got going again and von Rundstedt advanced to Kiev, which was defended in strength by the Soviets. In anticipation of this, he decided to deploy the majority of his armoured units south of the city, deeper into the hinterland.

On 12 July, the repositioning of the units was completed and von Kleist could begin his advance with the armoured units of Heeresgruppe Süd. He managed to drive the Soviets back across the Berdichev-Kazatin railway line and in this way disrupted their north-south connections. In response, the Soviets divided their forces into three: one part was positioned on the flank on the Dnieper near Kiev and another near Odessa, while a third retreated eastwards into a bend of the Dnieper. This created an inviting eastward bulge for the Germans. Von Rundstedt now faced a choice: either try to take the heavily defended city of Kiev – which would probably require some time – or advance eastwards to engage the Soviets there, with the risk that his left flank would be attacked by the Soviet units near Kiev. In any case, it was clear that a major confrontation would take place in this area, and its first move was an offensive by Romanian units across the Bug River on 3 August. They made contact with von Rundstedt's forces and encircled about 100,000 Soviets at Uman. This dealt a serious blow to the Soviet defence in the southern area of the Dnieper, and a second blow fell in September when the Panzerarmee Guderian advanced from the north in a wide movement to make contact with von Rundstedt.

However, the army leadership remained uncertain in the meantime about exactly what to do next. As far as it is possible reconstruct what happened, the following picture emerges of the differing views on the course to be followed. Von Bock, the commander-in-chief of Heeresgruppe Mitte, wanted to conquer Moscow as quickly as possible, regardless of the risks. Halder, Chief of Staff of the OKH, also wanted to continue the advance, but was concerned about the increasing influence of the OKW. Von Brauchitsch, the Commander-in-Chief of the Army, shared Halder's opinion but kept it to himself for fear of offending Hitler. Jodl, Chief of Staff of the OKW, was in favour of an advance on Moscow, but also took Hitler's side. Keitel, head of the OKW, lived up to his nickname 'Lakeitel' and agreed with the Führer. And Hitler himself, after a period of doubt, had finally come to the conclusion that the Ukraine front should take priority in the immediate future.

On 23 August, at a meeting at the headquarters of Heeresgruppe Mitte, a dejected Halder told those present, including Guderian, that the immediate goal in the coming period was the conquest of Ukraine and Crimea. Everyone reacted with shock to this decision, and Guderian was so critical of it that von Bock suggested that he and Halder fly back to Hitler's headquarters in Rastenburg, East Prussia. Von Bock assumed that Guderian, who was held in high esteem by Hitler, could change the Führer's mind. Halder and Guderian left for Rastenberg immediately and joined the conference held that same evening. Guderian was, however, warned by von Brauchitsch that it was not their intention to raise the issue of Moscow with the Führer. The operation to the south had been ordered, and the only question was how it should be carried out. On impulse, Guderian said that he might as well fly back to the front right away, but von Brauchitsch refused to give permission for this and ordered Guderian to report to Hitler on the state of the Panzerarmee, without however raising the Moscow question.

During the subsequent meeting, at which neither Halder nor Brauchitsch were present as representatives of the OKH, Guderian managed to bring up the subject of Moscow in a roundabout way and to present his point of view, which was militarily quite logical: Soviet resistance on the central front had been broken and an advance on Moscow would deliver a final blow, because the loss of the Russians' main rail, road and communications centre, their political 'solar plexus', would be a material and psychological blow from which they would not be able to recover. The operations in the south, on the other hand, would only yield local results. Hitler let Guderian finish and then began an explanation, the core of which was that his generals had no insight into the economic dimensions of war. He again emphasized the importance of the raw materials and agricultural production of Ukraine for the continuation of the war, as well the possession of the Black Sea coast, including Crimea, for the protection of supplies from the Romanian oil fields. For this reason he had given the order to deploy all available units to encircle Kiev and neutralize the Soviet forces there.

Guderian saw that he would achieve nothing by arguing with the Führer, especially since the rest of those present apparently agreed uncritically with their leader, a situation that would be repeated all too often in the future. Guderian had no choice other than to accept the change of strategy and return to his troops to make the necessary preparations for the upcoming offensive. Hoth, whose Panzergruppe had advanced even further towards Moscow, also had to focus on new objectives; he was to support von Leeb's offensive towards Leningrad. Only when the objectives in the north and south had been achieved would the two Panzergruppen be available again for an offensive towards Moscow by Heeresgruppe Mitte.

Von Bock and Halder were not entirely satisfied with the result achieved by Guderian, and Halder attacked him fiercely. Guderian attributed this outburst to the fact that Halder, in his opinion, was close to a nervous breakdown. Guderian replied correctly that it was not his job to raise such controversial issues with Hitler, since he was their subordinate. In early

South of Baturyn, Guderian is talking to Freiherr von Lüttwitz of the Schützen-Regiment 12 of the 4th Panzer Division. On the hood of the SdKfz 250 is a flag with a swastika as a signal to the Luftwaffe. In practice, this did not always work; in the rapidly changing battlefield, many German units were fired on by their own aircraft.

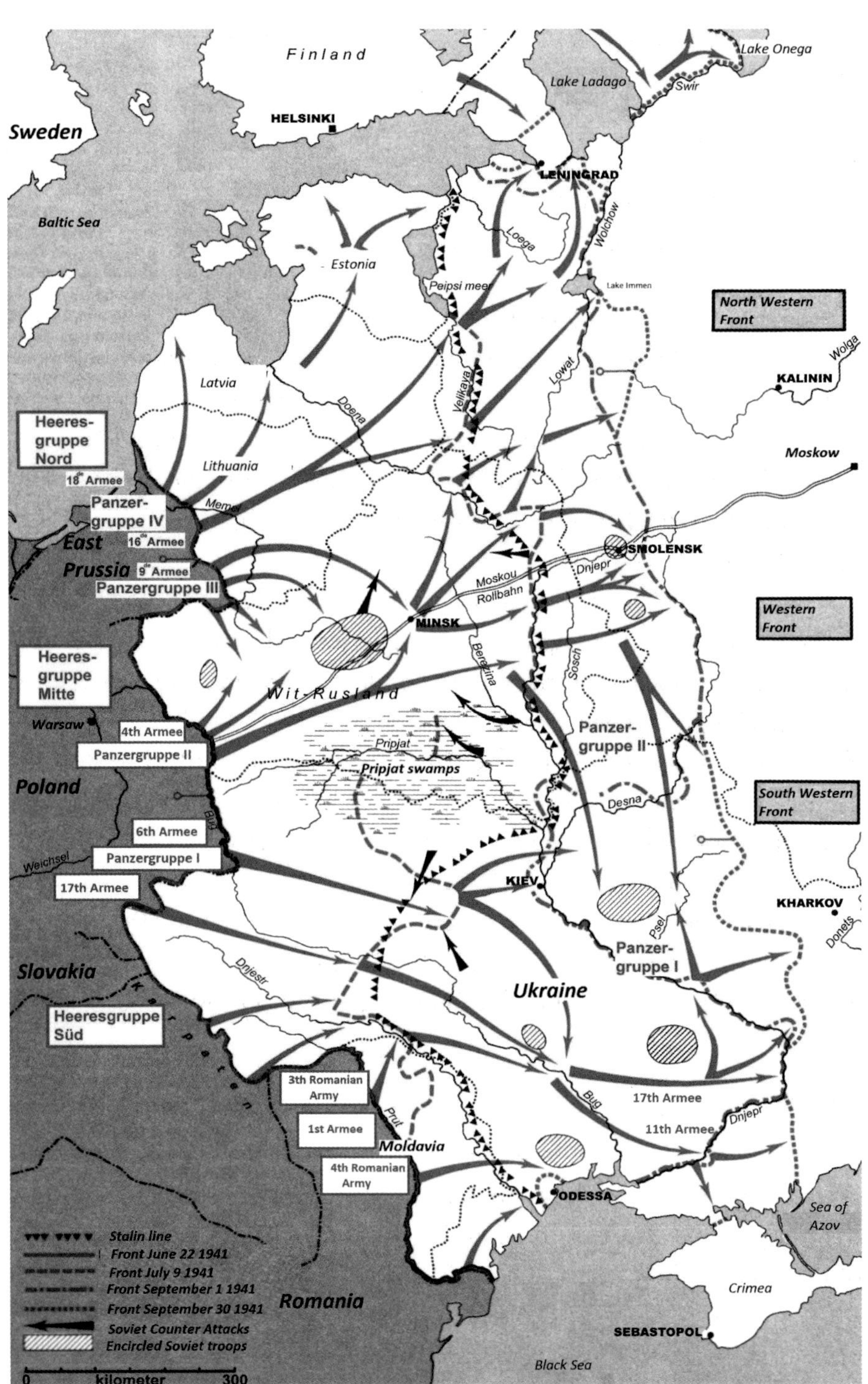

The advance of German troops in the period June–October 1941.

September, Guderian and his units began their south-westward march. His units advanced 350km to appear in the rear of the Soviets east of Kiev and complete the encirclement in the south. Guderian was aided in his advance by the misinterpretation of his movements by Stalin and the Soviet high command: they assumed that Guderian was merely making a southward flanking movement and would then turn north again towards Moscow. They reinforced their forces east of Guderian at the expense of those in the south. Guderian managed to fight his way south, and on 26 August his units crossed the Desna River; the road to Ukraine was open. But now Guderian's units were faced for the first time with the Russian *rasputitza*, the period of autumn rains that during the further advance in the Ukraine turned all roads into stretches of mud. Guderian recorded on 10 September that it took him ten hours to cover 160km; a day later, he needed even more time to cover only 120km.

After Guderian's units had crossed the Desna, von Kleist began the attack to the north, and on 16 September von Kleist's and Guderian's forces met at Sencha; the pincer movement was complete. In the area east of Kiev over 665,000 men, five Soviet armies, were encircled, and Guderian's and von Kleist's units were occupied in clearing the entire area until 23 September. Guderian describes the victory as a great tactical success, but the question now was whether there would be enough time before winter set in to take the important strategic goal of Moscow.

Hitler assumed that the Soviets were now no longer able to offer effective resistance anywhere on the front. He decided that the time was ripe to take Moscow, and in line with this, Guderian's Panzergruppe swung north again.

Crimea

In addition to Ukraine, Crimea had been the other major objective in this period. The Eleventh Army under its new commander, von Manstein, operating on the southern flank of Heeresgruppe Süd, was given the task of conquering the peninsula. On 12 September, von Manstein had taken over command from von Schobert, who had been killed in a forced landing on a minefield. At the time, the Eleventh Army had already crossed the Dnieper on its way to Crimea.

The problem with Crimea was that the peninsula was only connected to the mainland by a narrow strip of land, which consisted mainly of marshes. The German units could only reach the peninsula via the existing roads and railways. In addition, the area leading to Crimea consisted mainly of steppe, where the infantry of the Eleventh Army, which did not have tanks, were extremely vulnerable. Von Schobert had already been confronted with this when he initially wanted to force access to Crimea via a rapid action: his infantry units could not find cover on the steppe, and the attack was repulsed. Then the Soviets had launched a counter-offensive with their Eighteenth Army, and the Eleventh Army was even in danger for a short time of being surrounded. Von Kleist came to their aid with his armoured units, which were available after completing the encirclement at Kiev, and attacked the Soviets in the flanks and rear, destroying most of the Eighteenth Army.

After these battles Hitler assumed that the final phase of the battle could begin: the conquest of Crimea, the Donets River and Moscow. There was no indication that the Soviet Union could survive more than a few months more of fighting.

Taifun: the attack on Moscow

Guderian's Panzergruppe had meanwhile swung north again to prepare for the attack on Moscow; Operation Taifun. The units deployed for the attack on the capital consisted of:

- Panzergruppe II under Guderian
- Panzergruppe III under Hoth
- Panzergruppe IV under Hoepner, from Heeresgruppe Nord.

The plan was based on a double encirclement and was structured as follows: Hoepner and Hoth would encircle the Soviet units by meeting each other at Vjazma, while Guderian would first advance to Orel and then swing north to close the second ring of encirclement at Bryansk, together with Hoepner. In this way, the Soviets would be surrounded in two places and resistance around Moscow would be broken. Hoth would then advance from the north, Hoepner in the centre and Guderian from the south, to take the city.

Guderian's units took up their positions in the north on 30 September, and on 2 October Taifun began with a joint attack by the three Panzergruppen. On paper, Guderian should have received 100 new tanks, but only fifty had actually arrived at the front. The other fifty had been mistakenly transported to Orsha and did not arrive in time for the offensive. The fuel situation was dire, a harbinger of problems to come. The weather was still on the Germans' side, but new autumn rains could break out at any moment. In addition, the Soviets had used the time when Guderian was operating further south to strengthen the defences of Moscow.

Another factor that was beginning to play an increasingly important role was wear and tear on the panzers, especially those of Guderian's units that had already covered many thousands of kilometres. The previous campaigns, in Poland and the West, had been relatively short, both in time and in distance. Apart from the fact that the number of available tanks was steadily dropping due to losses, the field workshops had to expend ever greater efforts to keep the remainder operational. During this period, Guderian also noticed that his officers and men were increasingly becoming more mentally than physically exhausted: the months of almost continuous fighting had taken a heavy toll on them. This was in stark contrast to the mood at the OKH and in the high command of Heeresgruppe Mitte, where a quick victory was expected. Guderian noticed a growing gap between the reality of the front and the attitude of the high command. This gap would grow in the coming period and eventually become unbridgeable.

However, on 3 October Guderian's tanks again managed to break through the defensive positions of the Soviet Thirteenth Army without much difficulty and take Orel. With this, Guderian had gained control of an important railway and road junction, which formed an excellent base for future operations. The Soviets were so surprised by the rapid advance of the Germans that the trams were still running in the city when the first panzers appeared on the streets. The carefully prepared dismantling of industrial installations was thwarted by the rapid capture of the city, and the Germans found machine parts, crates of tools and semi-finished products along the streets leading to the factories. The captured airfield also offered the possibility of flying in supplies, and on 5 October Guderian contacted

Luftflotte II to transport 500 tons of petrol to remedy the Panzergruppe's dire fuel situation. The following day, Guderian was confronted with the increasingly active Soviet air force. Sevsk airfield was bombed, as was the local headquarters where Guderian was due to attend a meeting. While driving his car to visit one of his units in the afternoon, bombs fell on the left and right of the road, dropped by groups of three to six Soviet planes. They were flying too high to aim accurately, but Guderian immediately contacted Luftflotte II to urge better air protection in the coming period.

Meanwhile, the tanks continued their advance north and crossed the Bryansk-Orel road. Yeremenko, the commander of the Soviet forces around Moscow, realized the danger of encirclement and asked Stalin for permission

Recovery of tanks during battle: roadside assistance on the battlefield
During the Polish campaign and in May 1940, towing and repairing stranded tanks was relatively easy. After the battle was over, damaged tanks were collected and transported by rail to workshops in Germany. In North Africa and the Soviet Union, things were completely different; it was necessary to conduct maintenance on the spot. Towing tanks off the battlefield became a critical task, and this developed into a specialism. The Germans had special *Abschlepp Abteilungen* for this, using their SdKfz 9 FAMOs and other vehicles. The SdKfz FAMOs were the heaviest members of the SdKfz family, weighing 18 tons and designed to tow 21cm and 24cm howitzers. A single SdKfz FAMO could tow a PzKpfw III or IV tank, but for heavier tanks such as the Panther and Tiger, three FAMOs were needed. These were connected with 5m steel cables and then towed away, with the commander of the recovery unit in the leading vehicle.

These units operated in close cooperation with the armoured units they were supporting, often under enemy fire. They were also active at night, and all useful materiel, including the enemy's, was recovered. The commitment and quality of these units, and the workshops further behind the lines, made it possible for the Germans to compensate to some extent for the superiority of their opponents' tanks by getting many of their own damaged tanks ready for battle again in a short time.

to withdraw. This was refused, and on 6 October, against all expectations, the Germans took Bryansk, another important railway junction.

That same day, however, the 4th Panzer Division was confronted for the second time by the superior T-34s. They suffered heavy losses and were only able to hold out with difficulty. The T-34s were immune to the 37mm anti-tank guns and could only be knocked out, under particularly favourable circumstances, by the PzKpfw IV with its short 75mm gun. This had to manoeuvre behind the T-34 and hit the cooling grille above the engine compartment – a challenge for even the most experienced tank crew. What also concerned Guderian was that the Soviets were learning quickly: they had the infantry attack frontally and the tanks on the flanks, making it even more difficult to approach the T-34s from behind. Ultimately, the 88mm

Flak gun would be the weapon to take on these heavily armoured tanks. The Germans were also successful in the north: Hoth's units managed to capture Vjazma on 7 October, thus successfully completing the first phase of Taifun. During the past week, they had captured no fewer than 663,000 Soviet troops.

Mud made the already small number of roads completely impassable. Here an SdKfz 8 with a howitzer helps a horse and cart out of the quagmire.

Up to now, the weather had been kind to the Germans. But during the night of 6/7 October it started to snow. The next morning, a thaw set in and the fields and roads turned into one big mud puddle. Although the mud came too late to hinder the major encirclements of the previous week, it stopped the Germans from immediately switching to phase two of Taifun: the attack on Moscow itself. The plan was that this would be a classic pursuit of Soviet units towards the city, after which it would be taken by storm. In reality, the advance came to a standstill in the mud. Guderian recounted in his memoirs that the mud was central to the events of the following weeks. Trucks could no longer get anywhere without the help of tracked vehicles. The latter had never been designed to tow, and the wear and tear on engines, gearboxes and other parts of the drive train was therefore enormous. Since there were no towing cables available, ropes were sometimes dropped by aircraft to vehicles stuck in the mud. Some units could only be supplied by air for weeks.

The Soviets launched another attack on 11 October, this time with large numbers of T-34s, which inflicted heavy losses on German armoured units. Guderian decided immediately to draw up a report on this new threat. In it he described in the ways in which the T-34 was superior to the PzKpfw IV and what conclusions could be drawn from this regarding future tank design and production. He concluded that a commission should be sent to look at his sector of the front as soon as possible, and its personnel should consist of representatives of the Heeres-Waffenamt and the designers and manufacturers of tanks. At the front, this commission would not only be able to study tanks that had been knocked out, but also hear from the men themselves what should be included in the design of new ones. Guderian also urged the development of an anti-tank gun with sufficient power to destroy the T-34. On 20 November, the commission proposed by Guderian visited the front, and the conclusions of Guderian and the commission formed the basis for the later designs of the Tiger and Panther tank.

In addition to the change in the weather, another important event took place at this time: Timoshenko was replaced by Zhukov. This made little impression on the Germans; in their eyes, this was yet another general they had to defeat. However, history would teach otherwise; Zhukov would become the architect of the final victory over Hitler's Germany.

But this lay in the future. Now, Moscow was preparing for a large-scale evacuation. Ministries and other government organizations began burning important documents, and Lenin's coffin was moved from his mausoleum to a safer place. In the meantime, Zhukov's confidence increased when he saw that his flanks were able to hold their own against the attacks of Hoth from the north and Guderian from the south. He even foresaw the possibility of a flanking attack and possibly encirclement of Hoepner's units if the latter ventured too far east in the centre. On 14 October, Hoth managed to take one of the important strategic sites in the north, Kalinin, while Guderian tried to capture the important city of Tula, 150km south of Moscow. But Guderian found himself fighting the mud more than the enemy. On 29 October he was only 3km from Tula when his advance finally came to a halt. The Soviets put up tough resistance, and Guderian's supply line had been completely disrupted by the mud, leaving his units short of everything.

Slowly but surely, all along the front, the German advance ground to a halt. The Soviets, granted a respite, were able to bring in reinforcements. On 7 November, the chief medical officer reported to Guderian the first serious cases of frostbite, and by 13 November the temperature had dropped to 13° below zero. Guderian's men, unless they had been lucky enough to capture a Soviet uniform, were still wearing their cotton summer trousers. Guderian wryly noted in his memoirs that it was not until 30 August that the OKH had begun to realize that there might be a need for winter clothing in the months to come. The OKH report of that date noted:

> In the light of recent developments which may lead to actions against limited targets even in the winter months, the Operations Directorate will prepare a report on the winter clothing which will be needed for this purpose. After approval by the Commander-in-Chief, this report will be forwarded to the Operations Department for further action.

The Luftwaffe and the Waffen SS had already become aware of the reality of the situation: they had considerable quantities of good winter clothing available in good time.

On the morning of 14 November, Guderian visited units of the 167th Infantry Division. From conversations with officers and men, it became clear how dire the supply situation was. Winter boots, waterproofing for

Ever onwards. German tanks and motor vehicles covered great distances over dusty roads and difficult terrain. The wear and tear on all vehicles was unprecedented, and the number of deployable tanks fell continually. The tanks in the photo have spare chains mounted on the front as extra armour.

boots, underclothing and above all woollen over-trousers were all completely unavailable. In the afternoon Guderian visited the 112th Infantry Division, where he was told the same story. The men had been able to get their hands on Soviet overcoats and fur hats; only the national emblem they wore showed that they were German units. Although all their winter supplies had been sent to the front, these were little more than a drop in the ocean.

The number of operational tanks had also fallen to an unprecedented low: the three armoured divisions had only fifty at their disposal, whereas their nominal strength was six hundred. The telescopic sighting equipment of those tanks still operational had frozen because the grease which should have prevented this had not yet arrived. The units had no antifreeze. To start the panzers, fires were lit under them to warm the engine oil which had become viscous in the cold. Sometimes even the petrol froze.

As expected, Hitler now decided to make a new attempt to take Moscow. The plan was once again very ambitious: Guderian was to advance from Tula to Gorky, no less than 375km (!) east of Moscow. Hoth was to form the other arm of the pincer from the north, while Hoepner was to advance in the centre. On 19 November the attack began, and in response the Soviets threw everything they had into the battle, including a cavalry unit that charged the advancing Germans. On 30 November an engineers' assault unit reached positions 15km from the outskirts of Moscow, but this was the closest the Germans would ever come to the city.

Guderian struggled at Tula, which he had surrounded on three sides but was unable to take. On 23 November he told von Bock that he saw no chance of advancing any further. The OKH, which they called together, said that they understood the situation but that the orders had to be carried out. The next day, Guderian sent a liaison officer from his staff to the OKH to explain the difficulties, but his plea was turned down. Over the next three weeks Guderian repelled attack after attack by the Soviets, occasionally being forced to give ground. In the meantime, he continually asked von Bock to be allowed to withdraw to a line that was more defensible. After initially raising this with von Bock, he approached Schmidt, Hitler's adjutant, who arranged for Guderian to discuss the matter with the Führer himself. On 17 December he was given permission to fly to Rastenberg to report personally to Hitler.

The meeting took place on 20 December, lasted five hours and, according to Guderian, ended in complete failure. Guderian's memoirs are revealing in this regard. First, Hitler forbade him to give up even a metre of ground, indicating that the units should simply dig in to defend the territory they had taken. When Guderian pointed out that the ground in most places was frozen to a depth of one and a half metres and that digging was therefore impossible, Hitler suggested that holes should be shot by heavy howitzers, as was customary in Flanders during the First World War. Guderian replied that this was pointless because each of his divisions had only four howitzers with fifty shells, and these would only produce a few craters the size of a bathtub with a black rim around them. The cold was much worse than in Flanders, he added, and anyway, the shells were vital to resist Soviet attacks. When Hitler was asked whether he could provide sufficient alternative explosives, he merely repeated that not a single metre of ground was to be given up. Guderian said that a front line created in this way would cost the lives of many of his men, and that these losses would not only be pointless but also impossible to replace. Hitler then questioned the willingness of Guderian's men to die for their country; Guderian replied that his men were willing, but that they should be allowed a fair chance of survival, given that more casualties were being caused by frostbite than by enemy action. Hitler then declared that Guderian had lost sight of the bigger picture and identified too much with his men, who, according to his information, had received all

The cold is clearly visible in these photos. Due to the snow and ice, the Germans had to rely on sledges for logistical support. This required large numbers of men who were actually needed at the front.

the necessary supplies. However, after Hitler received more details in the days that followed, he had to admit that his information was incorrect. This led to a widespread campaign among the German population, directed by Goebbels, to collect warm clothing for the men at the front.

The next issue was transport and supply, which was insufficient to keep the units operational due to the great losses of vehicles in combat, the mud and the cold. Since no replacement vehicles were available, units had to fall back on sledges, which had a much smaller capacity and required a disproportionately larger number of men to deploy. Hitler demanded that fewer men should be involved in supply, thus freeing up extra troops to be deployed at the front. Guderian had great difficulty convincing Hitler that they were already using men and resources as efficiently as possible, and that his proposal would lead to a collapse of the supply system.

The subject then turned to accommodation. A few weeks earlier, the OKH had held an exhibition showing the way in which men were housed on the Eastern Front. It had been a great success, although none of the items exhibited had yet been put into production or reached the front. The continuous troop movements in the recent past had made it impossible to organize good accommodation, and the country itself offered few possibilities to shelter so many men. The picture that Guderian painted did not convince Hitler, but

This crew has dug in their Flak 88 gun to make it less conspicuous and give better protection against the biting wind.

it did impress Fritz Todt, who had joined the conversation. According to Guderian, Todt was moved by the account of life at the front and immediately offered two prototypes of stoves that had been specially developed for the Eastern Front. During dinner, Guderian related anecdotes about life at the front, but these did not interest Hitler. After dinner the meeting resumed, and in the course of conversation Guderian suggested that it would be wise to have officers with front-line experience fill positions on the general staff of the OKW and OKH, since the present campaign differed in every way from the First World War. The entire party, none of whom had been seen at the front for the last two years, reacted as if stung. Hitler immediately rejected the suggestion, and Guderian concluded that all his efforts had been in vain. The gulf between him and Hitler was complete and could not be bridged.

With a heavy heart, Guderian returned to his headquarters the next day. The situation there had changed dramatically: on 20 December, von Bock had been replaced by von Kluge, and Guderian knew that it was now only a matter of time before a definitive break would occur between him and the latter. After their umpteenth disagreement about withdrawal, von Kluge asked the OKH on 27 December to relieve Guderian of his command. Hitler agreed, and Guderian was placed in the officer reserve pool of the OKH until further notice.

On the run

Panzergruppe III described with shock the events in January 1942 on the road east of Klin:

> Discipline is disappearing. More and more soldiers are rushing west on foot without weapons. Some are leading a calf on a rope, others are pushing a sledge loaded with potatoes. The road is constantly being attacked from the air. Those killed by bombs are no longer buried. Men from support units (corps units, Luftwaffe, supply units) are fleeing in large masses to the rear.
>
> Without rations, freezing, irrationally seeking safety. Trucks, unwilling to wait at traffic jams, turn off the road and make their way across open fields and through settlements. The traffic control works day and night and is hardly able to keep traffic moving. The Panzergruppe cannot sink much lower than this.

In his memoirs, Guderian accused Hitler of shortsightedness, but Hitler's position on holding on to conquered territory was not entirely without sense. He feared that a retreat could easily turn into a flight, as happened, for example, in January 1942, when the Soviets launched a counter-offensive. Guderian's sacred belief in the superiority of mobile warfare, even in defensive actions, was now impossible to implement, since the German units were virtually immobile due to lack of transport and proper winter equipment. In addition, defensive action by armoured units had never been the subject of study or practice. If the Germans had attempted to disengage under these conditions from an enemy better equipped for winter fighting, they would have suffered heavy losses and would have had to abandon their heavy

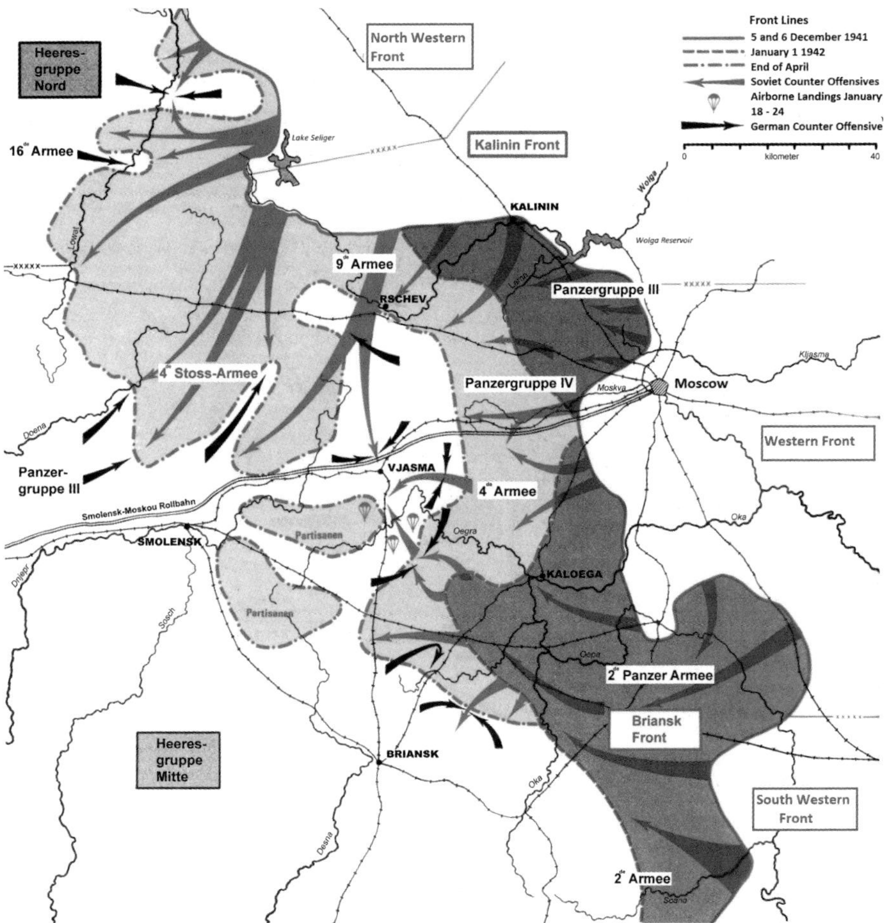

Soviet counter-attacks and the relief of Moscow.

weapons due to a lack of vehicles and fuel. If Guderian had had his way, his retreat would have ended in a wild rout, since his exhausted, poorly supplied units, ill-equipped to operate in winter conditions, were simply unable to carry out such a manoeuvre.

Hitler was right this time, but the 'hedgehog' positions that would be organized in the coming weeks proved to be a success that would have fatal consequences a year later at Stalingrad. Hitler's order to hold their ground now prevented the destruction of the 2nd Panzer Army, as was shown by its fortunes in the subsequent winter battles. Colonel General Rudolf Schmidt took over from Guderian and followed Hitler's orders. He was an obvious choice, since he had already managed several times to relieve trapped divisions with his aggressive defensive manoeuvres, and he was prepared to fight to the bitter end. Schmidt ordered that Guderian's units should withdraw into hedgehog defensive positions, usually built around one or more villages. Schmidt then achieved what Guderian had thought impossible, the creation of a new defensive front. The Soviets would batter themselves to death in the coming months on these stubbornly defended hedgehog positions, which proved to be an extremely efficient use of the available men and resources. By building these positions on the spot, large quantities of heavy weapons were saved from the Soviets, and these weapons, combined with clumsy Soviet tactics and the inexperience of the attackers, claimed many victims. Soviet units that managed to infiltrate between the various hedgehog positions were often quickly driven out again. The hedgehog position of the 116th Infantry Division at Sukhinichi, for example, held off two Soviet divisions for a month before being relieved by a counter-attack organized by Schmidt. In January, once the Soviet storm had subsided as their men tired, Hitler allowed Schmidt to make a tactical withdrawal in certain places, and by March he had built up a solid defensive line.

Crimea and Sevastopol

In the south, too, the German advance on Crimea had finally been halted. Von Manstein's plan was to advance as quickly as possible into the eastern part of the Kerch Peninsula, cross over to Kuban and from there continue the attack towards the Caucasus. On 18 October the attack began, and the Germans slowly but surely fought their way through the Soviet defences over the next few weeks. On 28 October, the day von Manstein intended to launch the final attack, the Soviets unexpectedly withdrew to Sevastopol, leaving the rest of Crimea to the Germans. They did not, however, give up Sevastopol and its fortifications, and von Manstein had no choice but to besiege it and draw up plans for its capture it in the coming months. The city remained a thorn in the German flank and a potential springboard for Soviet offensive operations.

Donets and Rostov

Meanwhile, the rest of Heeresgruppe Süd had not been idle. They had advanced further east to capture the Donets and Rostov area before winter set in. German units had managed to cross the Mius River between 12 and 17 October and had come within 50km of Rostov. If the Germans were to succeed in capturing Rostov, the road to the oil-rich areas of the Soviet Union would be open to them. But the weather changed, and Heeresgruppe Süd faced the same problems as the units before Moscow; as a result, the Red Army were given time to prepare the defence of Rostov.

Von Rundstedt succeeded in taking Rostov on 20 November, but the situation on his long northern flank became critical in the days that followed, and he was in danger of being cut off from the rest of Heeresgruppe Süd. He asked Hitler for permission to withdraw his units, but Hitler refused, because he simply could not imagine that the Soviets, after all the losses of the past period, would still be able to launch an offensive. He ordered von Rundstedt to hold on, but the latter ignored the order and began evacuating his units from the city. Hitler promptly dismissed him from his post and replaced him with Reichenau, the commander-in-chief of the Sixth Army, a confidant of the Führer.

The battle of Rostov

Reichenau quickly concluded that von Rundstedt had been right: the Germans would be unable to defend their advanced positions and their exposed long flanks against a determined Soviet counter-attack. On 1 December, barely a day after his appointment, he also asked Hitler for permission to withdraw behind the Mius River. This time, Hitler agreed, and the German units fell back in a coordinated manner to positions further west, significantly shortening the front line.

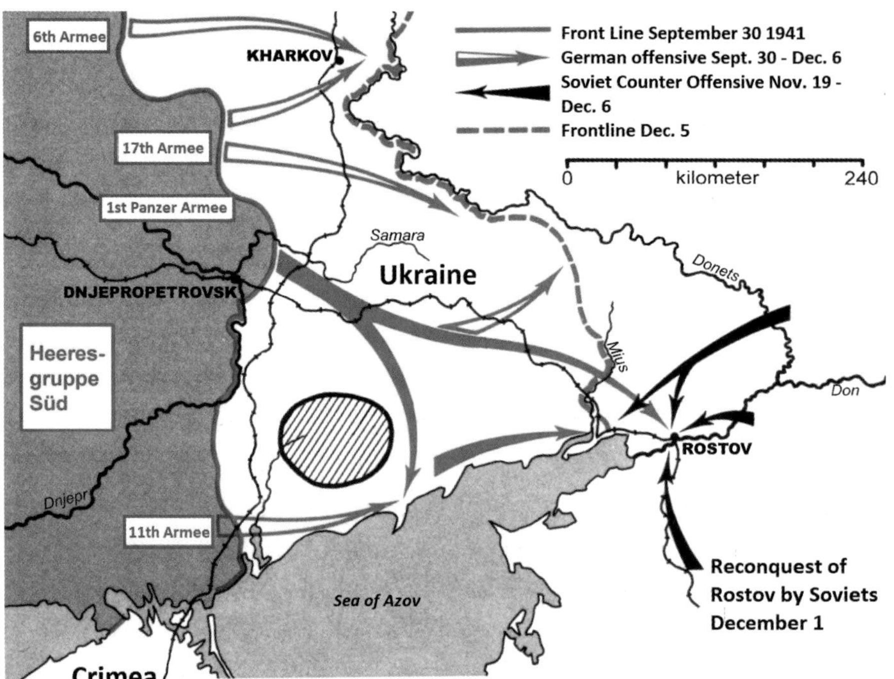

Fighting in the South.

And so the year 1941 ended. The Germans had achieved some of their objectives, but Leningrad and Moscow were still in Soviet hands. The failures of Barbarossa had cost a number of leading officers their positions, but Hitler was under the impression that the Soviet armies had been dealt a number of decisive blows. The winter considerably hampered the German armies, but they prepared confidently for what they saw as the decisive round in the coming months. No one could have predicted that the Red Army would soon take the initiative.

For Guderian, the war was over for the time being. However, he was in good company: at the end of 1941, a total of thirty-five army and divisional commanders had been dismissed. Von Brauchitsch, commander-in-chief of the army, was the most prominent victim. His position was now taken over by Hitler himself, who thus became both the titular and executive head of the army. Von Bock of Heeresgruppe Mitte, as we have seen, had to make way for von Kluge in December, von Leeb of Heeresgruppe Nord was dismissed in January, and von Rundstedt of Heeresgruppe Süd was transferred to the West.

Chapter 10

Guderian as Inspektor-General der Panzertruppen

1942

Before we continue to examine Guderian's career, we will first consider the year 1942, one of highs and lows for the Germans on the Eastern front. In December 1941 and the first months of 1942 the Wehrmacht was confronted with a series of Soviet counter-offensives, which sometimes pushed them back hundreds of kilometres. It was clear that the power of the Red Army had not been broken. However, the Soviets were unable to reach their objectives, a number of strategically important cities. In the spring, the Germans improved their position at various points on the front, with a view to mounting their own offensive, *Fall Blau*, aimed at capturing the oil fields in the Caucasus and finally neutralizing the Red Army. *Fall Blau* mainly involved the units of Heeresgruppe Süd, while the other Heeresgruppen had to hold the front; only in the Leningrad region was the offensive against the city to be continued. *Fall Blau* consisted of two parts: a southward and an eastward offensive. The southern offensive was aimed at the Caucasus and would be covered on the flank in the north by the offensive directed to the east. This was to create a protective front from Voronezh to Stalingrad, behind which the offensive towards the Caucasus could take place. Rostov formed a strategically important pivot for the whole operation.

The Germans and their allies could field 5,000,000 men with 1,400 tanks, the Red Army, 4,000,000 men and 2,000 tanks. The Soviets had managed to rebuild their arms production far to the east, and the first tanks from the new factories had rolled off the assembly lines in early 1942. *Fall Blau* got off to a successful start, and German units were soon deep in the Caucasus in the south and near Stalingrad in the east. Then, in August, the spearheads of both offensives were halted by stubborn Soviet resistance, and through a series of counter-offensives in December, the Soviets managed to isolate the German Sixth Army in Stalingrad. They also came close to cutting the lifeline of the German units in the Caucasus. The latter retreated quickly and managed to avoid being isolated. However, the Germans failed to relieve the encircled units at Stalingrad, and the entire Sixth Army was forced to surrender. Both sides had suffered heavy losses, but while the Germans would have great difficulty in bringing their armies back to an acceptable level in 1943, the strength of the Soviets continued to increase.

While all this was happening, Guderian had been living a secluded existence. His health had suffered greatly from the strain of the past few years, and his heart in particular had been seriously weakened. After a short rest cure in southern Germany in March 1942, he returned to Berlin, where his wife had been bedridden for several months with blood poisoning. Because of the constant stream of visitors to the Guderians in Berlin, he considered buying land in southern Germany, to which he could eventually retire. When Rommel, during sick leave, proposed Guderian as his replacement in North Africa, this was rejected, and it became clear to Guderian that he would probably never hold a command again. That decided his plans to settle in the countryside. He suffered increasingly from heart problems, resulting in a complete collapse in November, from which he recovered only slowly. By Christmas he was able to get out of bed for a few hours a day, and by the end of January he was more or less back on his feet. His idea was now to find a piece of land in Warthegau in West Prussia. This would be in the form of a gift from the Führer, given to those who had received the Oak Leaves to their Iron Cross. Eventually, he chose Deipenhof, an estate of 2,500 acres on excellent agricultural land. Guderian bought a herd of cattle and prepared for a life in the country.

In his memoirs, Guderian passed quickly over this period. However, there is another side to the story. He had already been receiving 2,000 Reichsmarks per month tax-free since August 1940, after he had been promoted to colonel general upon successfully completing the campaign in the West. This doubled his income. These special 'honourable allowances' were created to bind senior officers to Hitler and the Nazi regime. They were paid from a secret fund of the Chancellery and, according to Hitler, served to overcome the 'inner resistance' of officers so that they would conform to the wishes of the regime. The payments were continued in 1942, which shows that Hitler still envisaged a future role for Guderian. It was clear, however, that Guderian did not intend to make these payments public and that Hitler could stop them at any time. The acquisition of Deipenhof must also be viewed critically in this context. After having extensively explored the possibilities in Warthegau in the summer of 1942, Guderian had initially set his sights on another estate, which the secretariat of the Chancellery considered was too large and would set an undesirable precedent. Deipenhof was acceptable to them, which is why Guderian ultimately chose it. This 'gift' from Hitler was worth 1.25 million Reichsmarks.

Guderian returns

Although Guderian had fully prepared himself for civilian life, he received to his surprise an order from Hitler to report to the Führer's headquarters in Vinnitsa on 17 February 1943. Hitler had decided that the setbacks at Stalingrad and in North Africa called for a man of his proven calibre, something Schmidt and others had been urging. In addition, the General Staff and the Reichsministerium für Bewaffnung und Munition, responsible for the development of new tanks, were at odds with each other over future production policy. The men of the armoured units, in turn, had lost confidence in their leaders. In the middle of this decisive period for the Panzerwaffe, Guderian was appointed on 1 March to the position of Inspektor-General der Panzertruppen, with far-reaching powers over management, organization, equipment and training. His functional rank was that of army commander, answerable only to Hitler and having to consult only Zeitler, Chief of the General Staff. His authority extended not only to all tank and mechanized infantry units, but also to those of the Luftwaffe and the Waffen SS. The only exception to this were the armoured units of the Sturmartillerie, which deployed the formidable StuG assault guns. These were precisely the units that Lutz and Guderian had wanted to incorporate into the armoured divisions in the 1930s. The reason these units escaped Guderian's influence once again was that on a tactical level they did their work so well that outside influence was considered to be undesirable. They therefore remained under the command of the artillery. Ironically enough, these units would increasingly take over the role of the panzers in the future, because there would be an increasing shortage of tanks.

The organization and working methods of the Inspektorat-General der Panzertruppen

After being appointed, Guderian chose as his chief of staff Colonel Thomale, an officer with great experience at the front. Guderian also had the services of two officers from the General Staff, one for the organizational side and one for personnel matters. The former was Lieutenant Colonel Freyer, who was no longer fit for front-line service due to serious wounds; the latter was Major Kauffmann, later replaced by Major Freiherr von Wöllwarth. Guderian's adjutant was Lieutenant Colonel Prinz Max zu Waldeck, who had also been gravely wounded. The rest of the staff was made up of older officers with considerable combat experience but who, due to wounds, were not eligible for anything more than short periods of active service. These men were rotated regularly so that, in Guderian's words, 'they could exchange the dusty atmosphere of an office for the fresh air of the front'. Regular rotation ensured that the Inspektorat-General remained in close contact with the realities of combat. In addition, a number of mens were appointed as liaison officers between the Inspektorat-General and front-line units. These were also generally officers who were recovering from wounds and were not yet fit for active service. They studied and evaluated the *Erfahrungsberichte* from the front and investigated unexpected events that had occurred during the fighting. Their evaluations were processed into tactical instructions for the combat units and served as learning material for the various training courses undertaken by armoured units. These courses were combined with those of other army units, so that everyone became aware of the most recent lessons.

In order to get a sense of the political-military landscape, Guderian had a conversation with Feld Marschall Milch at the Reichsluftfahrtministerium (Ministry of Aviation). Milch indicated that only a limited number of people actually had any influence on Hitler, namely Goebbels, Himmler and Speer. On 3 March, Guderian visited Goebbels and they discussed the current leadership structure of the Wehrmacht. According to Guderian, too many institutions and interests were involved: the OKW, the OKH, the Luftwaffe, the Waffen SS, the Kriegsmarine, the Reichsministerium für Bewaffnung und Munition, each with its own preoccupations. All these parties tried to put their case with Hitler, which regularly made the decision-making process opaque and even threatened to render it chaotic. Guderian proposed that Hitler be supported by an officer at the level of the General Staff with substantial recent front-line experience – in other words, someone unlike Keitel. Goebbels listened to everything attentively and undoubtedly formed his own opinion about this. He promised in any case to bring up the matter with Hitler if the opportunity arose.

During the same period, Guderian had his first conversation with Speer, his important counterpart in the field of tank production and the production of armoured vehicles. The subject of the discussion was the current situation of the Panzerwaffe in 1943. Due to the losses of the previous years, the strength of the Panzerwaffe had slowly but surely been eroded. As late as 1942, two new divisions had been established with the aim of increasing the fighting power on the Eastern Front: the 26th Division on 15 September and the 27th Division on 2 October. But in the first months of 1943, the 14th, 16th and 24th Panzer Divisions were lost at Stalingrad, and later in the spring in North Africa, so were the 10th, 15th and 21st. In addition, two weak divisions, the 22nd and 27th, were disbanded in the same period. All in all, in the space of a few months, the size of the Panzerwaffe had been reduced by eight divisions.

The Guderian Plan 1943–1945

Guderian was faced with the task of getting the Panzerwaffe back on its feet after the traumatic year of 1942. His main idea was that 1943 would be used to restore its fighting strength, and his entire plan was focused on this. First of all, he focused his attention on production. On closer inspection, it turned out that a discussion had taken place within the General Staff at the beginning of 1943, which almost led to tank production being limited to Tigers and Panthers, the latter being still in the prototype phase. If this line had been followed, total German tank production would have been just twenty-five Tigers per month. In the autumn of 1942, production of the PzKpfw III had already been stopped, and only chassis of this tank were produced which served as a platform for various anti-tank and artillery guns. In addition, in order to free up production capacity for the new medium-weight Panther, halting production of the PzKpfw IV was also considered, again with the exception of its chassis. The General Staff had reversed its decision in time, but it was illustrative of the gaps that had arisen between the reality of the battlefield, the views of the high command and the production capability of German industry. The PzKpfw IV remained in production until the end of the war and would form the backbone of German armoured units, together with the many surviving examples of the PzKpfw III. Tanks such as the Tiger and the Panther, however impressive their performance was, were only able to achieve local superiority. They had little influence on the outcome of the war.

On 9 March, barely a week after taking office, Guderian laid out his plans for the Panzerwaffe in 1943/1944 to Hitler. The location was Vinnitsa, and to Guderian's surprise, not only was Hitler there, but the entire OKW, the Chief of the General Staff, the Inspectors General of Infantry and Artillery and Schmidt. They were not only very interested in Guderian's plans, but had also studied a résumé that Guderian had previously sent to Hitler and had prepared themselves thoroughly for the coming discussion. We will look at these plans in broad outline below.

Goals for 1943

The chief goal for 1943 was to form a limited number of panzer divisions as efficient fighting units capable of conducting limited offensives; by 1944, panzer divisions across the board should be able to conduct large-scale offensives again. Guderian argued that a panzer division was only an efficient fighting unit if other weapons and vehicles within the division were realistically related to the number of tanks available. German panzer divisions should be built around four battalions of tanks, leading to a total tank strength of 400 per division. If the number of tanks was less than 400, the total complement (men and vehicles) of the division would no longer be in proportion to its offensive capacity. The aim was to prevent the creation of a surplus of support units, with large numbers of vehicles and men, which would become a drag on both command and supply. At the time Guderian wrote this, almost all divisions had a nominal strength of fewer than 400 tanks. As a result, it made more sense to have a limited number of fully equipped divisions than a large number of partially equipped ones with a disproportionate number of support units. Based on this view, Guderian continually advocated in the following months that panzer divisions that had suffered too many losses should be withdrawn from the front to be re-manned and re-equipped.

Production of tanks

The PzKpfw IV was still the most important German tank and would remain so until the end of the war. In Guderian's eyes, all efforts should be focused on increasing production of the PzKpfw IV F2 with the 75mm L/43 or L/48 anti-tank gun, so that each month, a tank battalion could be completely re-equipped with this model. During 1943, a limited number of battalions could be equipped with the new Panthers and Tigers. The first Panther battalions could not be operational until July or August at the earliest, given the many teething problems that still plagued this tank. In order to maintain combat strength in the coming period, Guderian proposed equipping the panzer divisions with light assault guns, which were now rolling off the assembly line in large numbers. These consisted of a 75mm/L 48 gun on a PzKpfw IV chassis. Guderian assumed that it was possible to equip one battalion per month with this gun, and this process should continue until industry was able to produce enough tanks to equip all panzer divisions with tanks. Production of the PzKpfw IV should be further increased in 1944 and 1945, but this increase in production should not be at the expense of the production of the Tiger and Panther.

In addition, tanks' operational life had to be extended. Guderian had the following measures in mind:

- Thorough testing and perfection of new tank types such as the Panther
- Thorough training of tank crews at individual and unit level; crews also had to be involved in the delivery of tanks from the factories and the application of the latest modifications
- Allocating sufficient training materials to the training units
- Ensuring continuity in training and preventing units from being deployed prematurely at the front.

The underlying reason for these measures was that many tanks suffered unnecessary damage due to injudicious operation by their crews.

Guderian's later booklets, *Pantherfibel* and *Tigerfibel*, also served to give crews instruction about what tanks could and could not do. For example, attention was drawn to the vulnerability of the drive train (gearbox, drive shafts, etc.) and the fact that driving a tank through buildings could cause serious damage to the tank itself.

Deployment of tanks

According to Guderian, the success of tank divisions depended essentially on the concentration of sufficient tanks at the right place in the front (the familiar *Schwerpunkt*). For this, sufficient tanks had to be available in the most important places. To achieve this, Guderian envisaged the following measures:

- Fronts of secondary importance were no longer to receive new tanks; they should limit themselves to captured ones
- Tank battalions (PzKpfw IV, Panther and Tiger tanks) were to be concentrated within armoured divisions commanded by officers experienced in mobile warfare
- New equipment such as the Tiger and Panther tanks and the new assault guns were only to be deployed if they were available at the front in sufficient quantities to make optimum use of the surprise effect of these new weapons. The premature deployment of new equipment simply invited the enemy to take counter-measures in the form of new weapons that could be deployed within a year, if not sooner. The experience of the premature deployment of Tiger tanks at Leningrad had demonstrated this danger
- It was necessary to avoid the creation of new divisions. Far better to rebuild badly damaged armoured and motorized divisions around the remaining cadres and men, because they had thorough knowledge and experience of the equipment and were valuable in rebuilding their division. Newly created units could never match this knowledge and experience
- The existing policy of letting armoured divisions burn out slowly but surely at the front prevented them from being regrouped and

re-equipped in time for future offensive actions. In line with this, a large number of armoured divisions would have to be withdrawn from the front.

Anti-tank guns

Defence against tanks would increasingly rest on the shoulders of the assault guns, as all other anti-tank guns proved to be less and less effective against the newly deployed Soviet tanks. All divisions on the main fronts were to be equipped with assault guns. Secondary fronts were to build up central reserves of the weapon, while their divisions were to be equipped with mechanized anti-tank guns in the near future. Slowly but surely, the existing anti-tank battalions would be replaced by battalions of assault guns. The new heavy assault guns were to be used only on the main fronts and for special missions. Their role was to destroy enemy tanks. Furthermore, infantry divisions were to be equipped with a newly developed type of light assault gun, because their anti-tank guns towed by half-tracks were too vulnerable.

Armoured reconnaissance battalions

These units had become the neglected element of the armoured divisions. They had been of great importance in the first years of the war, in Europe and North Africa, but they played a less prominent role on the Eastern Front. If, according to Guderian's ideas, the Wehrmacht was to take the initiative again in 1944, there would be a growing need for the following types of reconnaissance vehicles:

- Light 1-ton armoured troop transport (half-track) vehicles, production of which had just started
- Armoured reconnaissance vehicles with sufficient armament and armour and a speed of between 50 and 70kph. Such vehicles had yet to be designed.

Panzergrenadiere

It was crucial that the existing 3-ton armoured half-tracks continued to be produced in sufficient quantities. The engineers and signal units had to use the same type of vehicle.

Artillery

The armoured and motorized divisions should receive sufficient numbers of mechanized guns in the near future; this was a wish that Guderian had cherished for ten years. Artillery observers had to have the latest models of tank at their disposal.

In summary, Guderian wanted the following main points to be agreed:

- The Inspektorat-General der Panzertruppen should be responsible for all assault guns (light and heavy)
- No new armoured or motorized divisions to be created, either for the Army or the Waffen SS
- The existing Waffen SS divisions and the Herman Göring Fallschirmjäger Division should be incorporated into the organization of the army

The Hummel was built on the chassis of a PzKpfw IV and equipped with a 150mm howitzer.

- Production of the PzKpfw IV should be continued during 1944 and 1945
- Further research to be conducted into the possibility of developing a new model of light assault gun for the infantry divisions.

All the above points gave rise to lively discussion, but those involved endorsed the main lines of Guderian's recommendations. The only contentious point was the placing of all assault guns under Guderian's responsibility; everyone except Speer was strongly opposed to this for various reasons. The discussion had far-reaching consequences: the assault guns remained an independent weapon, the anti-tank battalions retained their ineffective anti-tank guns pulled by half-tracks, and the infantry divisions were denied adequate anti-tank defence. Nine months later, Hitler would recognize that this was a major mistake, but it would never again be possible to provide all the divisions with the necessary weapons. After the group had argued for four hours, Guderian left the meeting and promptly fainted outside the door, so affected was he by the heated discussions.

The excellent StuG assault guns remained under the command of the artillery. In the coming years, the role of the StuG assault guns would become increasingly important due to the defensive nature of the battle. Here is a StuG III G with the long 75mm L/43 gun at full speed. Note the extra armour in the form of caterpillar tracks.

These additional demands placed a heavy burden on German industry; quite apart from relatively standard products, the production process was characterized by an almost infinite number of modifications and adjustments, resulting in an unclear and often unpredictable need for spare parts, with associated logistical nightmares. The fact that German industry did not collapse entirely in the face of these demands was due in no small part to Speer, who was appointed by Hitler in February 1942 as Minister of the Reichsministerium für Bewaffnung und Munition (from September 1943, Reichsministerium für Rüstung und Kriegsproduktion). Speer succeeded Fritz Todt, who had died in mysterious circumstances after a meeting with Hitler, and he initiated a policy characterized by:

- Standardization of weapons systems
- Simplification of weapons systems
- Rationalization of production processes.

He managed to introduce changes of all kinds to German industry, and production duly increased in leaps and bounds. In doing so, he had to overcome a great deal of resistance from the management of many German firms, who attempted to continue prioritizing quality rather than turning out the necessary quantity of equipment. Production increased spectacularly in 1943, partly due to the efforts of Speer, but also as a result of action taken by his predecessors in 1942. The focus was not only on manufacturing military equipment, but also on producing machinery to allow companies to expand their production capacity. Due to the availability of this new capacity, production in 1944 would show an even more impressive upward curve.

Production figures in 1943

Production of the PzKpfw III and the PzKpfw 38(t) was reduced and then stopped in 1943. The newcomer was the Panther tank, while production of the Tiger tank rose from 84 to 647 and production of the PzKpfw IV from 994 to 3,013, all tanks with the excellent 75mm L/43 or L/48 gun. Production of the StuG III and the new StuG IV rose from 789 to 3,042. Total tank production rose from 4,475 to 8,705, an increase of 94 per cent.

In the days following his first appearance, Guderian visited the various factories where tanks were produced, and on 19 March attended a presentation to Hitler of a number of new weapons, namely the 'Gustav' railway gun, the Ferdinand tank and the PzKpfw IV with *Schürzen* (aprons). These were hung loosely on the side of the PzKpfw III, IV and the assault guns and ensured that anti-tank grenades, particularly from the Soviet infantry, exploded before hitting the less heavily armoured horizontal surfaces and vital parts of the running gear of the tank itself. In practice, they proved to work very well. However, Guderian did not share Hitler's enthusiasm for the Ferdinand tank, developed by Porsche and named after him; it had a fixed 88mm L/70 gun but no additional armament in the form of machine guns, and could therefore not defend itself in close-quarters combat. Guderian considered this a major weakness, despite its impressive armour, effective gun and innovative electric drive. Since ninety had already been built, however, a useful application had to be found for it, and eventually, it was decided to

Guderian had little confidence in the Ferdinand tank, named after Ferdinand Porsche. His fears were confirmed during Operation Zitadelle (the German offensive at Kursk in 1943). An interesting novelty on this tank was its electric propulsion provided by generators powered by the petrol engine.

place these tanks in a special armoured regiment consisting of two battalions of forty-five tanks each.

The 'Gustav', an 800mm railway gun that required a double track to move, was not Guderian's concern until Hitler pointed out that, according to Müller, Krupp's chief engineer, this gun could be used against tank concentrations. All Guderian could say was that this might be so but it was unlikely that it could ever actually hit a tank. The initially furious Müller later had to admit that his claim had been unfounded.

After dealing with these weapons systems, Guderian turned his attention to the Herman Göring Fallschirmjäger Division, which had been living a leisurely life in the Netherlands for the past years. This unit, of no fewer than 34,000 men, was barely able to provide a single battle-ready division.

Kharkov

While this was happening in Germany, the fighting in the Soviet Union had continued unabated. Guderian was particularly interested in what was happening around Kharkov, which was back in German hands after a hard battle. On 29 March he flew to the headquarters of Heeresgruppe Süd to discuss recent operations with von Manstein. Guderian considered the events of the past months a textbook example of how armoured formations should be deployed and was particularly interested in the experiences of the Tiger battalions of SS divisions Gross Deutschland and Leibstandarte SS Adolf Hitler. His aim was to gain an idea of the technical and tactical capabilities of the Tiger tanks so that he could work out how they could best be organized and deployed in the future. Guderian also met his old comrade-in-arms Hoth, now commander of the Fourth Panzer Army, part of Heeresgruppe Süd under von Manstein. Guderian had great admiration for von Manstein and would later, in his role as chief of staff, advocate appointing him as head of the OKW in place of Keitel, whom he detested. However, von Manstein was an extremely outspoken individual who had got on the wrong side of Hitler. In the Führer's opinion, von Manstein could only function well when he had new, fresh divisions under him; moreover, he was poor at offensive action. In the following years, however, Von Manstein would show that he was an extremely capable commander, who could deal very competently with divisions that had emerged battered from battle.

Besides offering a good example of how tanks could be deployed, Kharkov also provided a fine example of the idiosyncrasy of the commander involved,

Peiper's armoured ambulances

Immediately after their arrival in late January, the units of the Waffen SS had spread out over the snowy plains to receive the men of the 298th and 320th Infantry Division and escort them to their lines. In a few cases, German infantrymen who had been captured were freed by Waffen SS reconnaissance units, who attacked Soviet positions on motorcycles with sidecars while firing with all their machine guns.

The units of the SS Division Leibstandarte were able to establish a corridor to survivors of the 320th Division in early February, but this was cut on 12 February. Encircled, and with thousands of wounded in their ranks, they had to be relieved as quickly as possible. A Kampfgruppe under the command of SS-Sturmbannführer Joachim Peiper, then the battalion commander of 2nd Panzergrenadiere Division, equipped with half-tracks, was ordered to rescue them. His battalion was given a unit of StuG IIIs and a column of ambulances. The Kampfgruppe managed to punch a hole in the Soviet lines, destroying several Soviet tanks and advancing 50 kilometres behind the Soviet lines to find the remnants of the division. After the wounded were hoisted into ambulances, Peiper's men returned to the German lines to discover that a unit of Soviet ski troops had cut off their route and destroyed the only bridge over the Udy River. The Waffen SS attacked the Soviets and, after house-to-house fighting, managed to capture and repair the bridge, whereupon the ambulance column was transferred under escort back to the German lines. However, the heavy StuGs and the half-track vehicles could not cross the temporarily repaired bridge, and Peiper had to find another crossing. After a search behind Soviet lines, another possible crossing point was found, and Peiper finally got his units and the rescued infantrymen back to the Leibstandarte's lines. His losses were no more than a few dozen killed and wounded.

Paul Hausser. The situation was as follows. After the fall of Stalingrad, the Soviets had advanced rapidly to the west and in early February they threatened to cross the Dnieper River and take Kharkov. Soviet units also advanced south to cut off German forces that were rapidly withdrawing from the Caucasus. The Germans brought in units from all over Europe to stabilize the front in the east. One of these was the 1st SS Panzer Corps under the command of Hausser. This newly formed SS Panzer Corps, consisting of the SS divisions Gross Deutschland and Leibstandarte SS Adolf Hitler, was transported from France to the Kharkov region in 200 trains in two weeks. The first units arrived at the end of January and established a defensive ring around the city. The Panzer Corps slowly built up its strength in the weeks that followed and, in addition to establishing a defensive line along the banks of the Donets, ensured that the survivors of the 298th and 320th Infantry Divisions were extracted successfully after a long journey west across the steppe. Waffen SS units regularly carried out deep operations on the east bank to relieve these units. The infantry occupied the defensive lines, while Hausser positioned his armoured units as a strategic reserve in the rear.

South of Kharkov, however, a gap of 160km had opened up between the right, southern flank of Hausser's units and the left, northern flank of the First Panzer Army operating in the south. The inevitable happened, and after fierce fighting, the Soviets managed to penetrate the gap on 14 February and almost completely encircle the city: a new Stalingrad loomed for Hausser's units. Despite urgent requests, Hitler refused to give permission for a retreat. On 15 February, Hausser took matters into his own hands and ordered his units to break out. Hitler reacted furiously, but Hausser avoided the usual court martial because the Führer eventually realized that Hausser had made the right decision and that he could not afford to lose the SS divisions.

During the same period, a number of Soviet units managed to advance deep into the no-man's-land that had been created. But the Germans were far from defeated. First, the leading elements of a Soviet tank corps were cut off from their supply units by the combined action of the 11th Panzer Division and the 333rd Infantry Division. Radio traffic indicated that the Soviets were running out of fuel, and in a classic example of mobile warfare, German armour pushed the Soviet units back, leaving the infantry to eliminate the stalled tanks.

On 19 February it was the turn of the Soviet Sixth Army, which had advanced far to the west. Hausser attacked its new positions, with his SS Panzer Corps on the northern flank, while the 48th Corps took care of the southern flank. The two met on 23 February at Pavlograd and cut off the furthest advanced Soviet elements. The commander of the Soviet Sixth Army was ordered to advance even further, because aerial reconnaissance indicated that the Germans were still retreating. On 28 February, the curtain fell on both the aforementioned Soviet tank corps and the Sixth Army. Both were surrounded by the Germans before they could reach the Donets. Although the number of prisoners of war was fewer than expected due to the encirclement being less than watertight, the majority of the Soviet materiel fell into German hands.

Von Manstein now smelled success and immediately sent Hausser's units north again towards Kharkov. The icy weather provided hard surfaces, which made a rapid advance possible. Hausser pushed aside the 3rd Soviet Tank Corps on his way to Kharkov and reached the city just before the thaw set in. On 8 March his units captured the access roads to Kharkov, cutting it off from the rest of the world. The Germans then entered the city and on 15 March were once again its masters, a month after Hausser had ignored Hitler's order to hold it at all costs.

The Germans now kept up the pace of their advance and moved north to recapture Kursk. The Soviet Sixty-Ninth Army, which was on Hausser's route but had already been badly damaged in previous battles, could not offer any serious resistance and was neutralized. On 17 March Hausser attacked Belgorod, taking the city the next day. After this, the German offensive came to a standstill. Mud and exhaustion caused both sides to pause. Hausser's units had paid a heavy price during the two-month campaign, suffering 11,500 dead, missing and wounded, of whom 4,500 were from Leibstandarte SS Adolf Hitler. The losses of equipment were also considerable: only half of the tanks with which they had arrived from France were still operational.

This, and Balck's actions further south, were examples of the type of mobile warfare that would become increasingly rare in the coming years. The Germans would be increasingly forced onto the defensive, and the new Tiger and Panther tanks would be used primarily to establish a temporary local superiority or to cover a retreat.

The end result of the fighting in the first months of 1943 was that there was a large bulge west of Kursk within which six Soviet armies were stationed. Von Mansein's plan was to cut these units off by having elements of Heeresgruppe Mitte attack from the north and his own units from the south. This laid the foundations for the largest tank battle ever, Operation Zitadelle.

The Waffen SS

Back in Germany, Guderian revived the question of the special position of the Waffen SS. In early April, he met with Himmler in Berchtesgaden to discuss this subject and advocated building these units in the same way as the Army's, and also restricting the growth of the Waffen SS as an institution. This last point in particular met with great resistance from Himmler and Hitler, who wanted to keep the Waffen SS as an independent entity. In their opinion, the Army could not always be trusted when push came to shove, and so the Führer and the NDSAP needed an army of their own. The Waffen SS would act as a Praetorian guard which would follow Hitler at all times if the army was no longer to be trusted. Because the Waffen SS did not become part of the army, it was identified after the war with the Allgemeine SS (the administrative branch) and the SD (security service). The division between the Waffen SS and the army grew during the war, mainly because the Waffen SS received the best and the most weapons. However, one must not forget that these units fought side by side on all fronts as brothers in arms, as was evident from events around Kharkov.

Opposition to Hitler

Hitler's fear of conspiracies or coups was not entirely unfounded, as several groups inside and outside the army actively plotted to put an end to the regime. For example, Guderian recorded in his memoirs that he was approached by Gördeler in February 1943. After a successful political career during the Weimar Republic, Gördeler was elected Mayor of Leipzig in 1930. He was a declared opponent of National Socialist racial ideology and left his party, the NVPD, in 1931 when it began to collaborate with the NSDAP. After 1933, Gördeler was one of the few politicians who opposed the

Nazis. He tried several times to help Jewish businessmen from Leipzig who were threatened by the economic policy of Aryanization. When the Nazis ordered the destruction of the monument to the German-Jewish composer Felix Mendelssohn in 1936, Gördeler tried to rebuild it. When he failed he resigned after his re-election as Mayor of Leipzig. Between 1937 and 1938 Gördeler spent a lot of time abroad, mostly in France, Great Britain, the US and Canada, warning anyone who would listen about what he saw as the aggressive and dangerous foreign policy of the Nazi regime. During this period, he met Winston Churchill several times. Gördeler was deeply disappointed by the British role in the Munich agreement in 1938. After Munich, he gathered conservative politicians and generals around him, drew up a new constitution with them and even made a list of names of potential ministers. Through his friend Dietrich Bonhoeffer, he also managed to involve the so-called 'Freiburg Kreis', a group of professors from the University of Freiburg, in the drafting of the constitution. The intention was that Gördeler would become chancellor after a possible coup.

Half-tracks in their classic role as tractors were indispensable in the difficult terrain of the Soviet Union. Here an SdKfz 7 is on the move with the barrel of a 150mm Schwere Feld Haubitze. The crew keeps a close eye on how everything fits and whether the bridge can handle the weight.

Gördeler discussed with Guderian ways in which Hitler's power could be limited. One step was for the Wehrmacht to tackle the party leadership, although it was unclear which commanders would support such an idea. Gördeler asked Guderian to find out during his visits to the front who he could count on to develop his ideas. When Guderian asked who was actually in charge of the whole idea, Gördeler named Beck, a commander whom Guderian described in his memoirs as an eternal doubter. Guderian indicated that he had little interest in the idea, but promised to put feelers out at the front. In the following period, none of the commanders seemed to display any sympathy for Gördeler's plans, partly because of the serious situation at the front, but also because of their oath of allegiance to Hitler and because the design of the whole thing did not appeal to them. Guderian reported this back to Gördeler in April, and Gördeler did not contact him again. He continued to plot, however, and was eventually arrested and shot by the Gestapo after the failed assassination attempt on Hitler in 1944.

North Africa

A different front, meanwhile, demanded Guderian's attention in early April: North Africa. While the situation in North Africa had become hopeless, new units were still being sent there, including the newest Tiger tank battalion, Schwere Panzer Abteilung 501. The new units had no chance of survival and would soon be lost. Guderian advocated evacuating all tank personnel, not only the crews but especially the maintenance technicians with their years of experience. These men could serve as the basis of new units to replace those lost in the East. But Hitler was not interested in this proposal, and valuable expertise was lost in the following months.

Despite their heavy losses, the Germans were able to rebuild a decently equipped panzer force by mid-1943, thanks to the efforts of industry. They had succeeded in replacing four of the eight broken divisions with new ones, built around the core of survivors of the divisions destroyed in 1942 and using men who had been earmarked for other divisions, combined with all the materiel that could be salvaged. As a result, the Germans were able to field thirty-two armoured divisions in the spring of 1943, compared to twenty-eight at the beginning of 1943. However, these divisions were largely below their nominal combat strength. Twenty-one divisions of this force, including four SS and two Panzer-Grenadiere Divisions, were to be deployed, much against Guderian's wishes, on the 1943 summer offensive in the Soviet Union, Operation Zitadelle. The fact that only one other panzer division, the 13th with seventy-one tanks, was deployed elsewhere on the Eastern Front illustrates the fighting power that the Germans were concentrating at Kursk. Four other tank divisions were at that time in Sicily and on the Italian mainland, five including three new SS divisions, were in France, and one was in the Balkans.

After a visit to France, Guderian was recalled to Berlin at the beginning of May to discuss Operation Zitadelle. During two sessions on 3 and 4 May, this offensive, conceived by Zeitzler, Chief of Staff of the General Staff of the OKH, was examined in depth. All the key figures were present: the members of the OKW, von Manstein of Heeresgruppe Süd, von Kluge of Heeresgruppe Mitte, Model of Heeresgruppe Nord, and Speer. Keitel assumed that despite the fact that the German units had suffered heavy losses in the preceding months, the initiative could be regained by deploying the new Tiger and Panther tanks. Model gave the meetimg the latest information on the Soviet position. The Red Army were building deep and very strong defence lines, with an abundance of anti-tank guns and artillery in precisely those areas where the Germans planned to force a breakthrough. They had also withdrawn most of their mobile units and positioned them behind the front in anticipation of the pincer movement planned by the Germans. Model's conclusions were that either a completely new plan had to be developed, or the entire offensive had to be abandoned. Guderian, asked for his opinion, reminded the meeeting that the units at the front had just been reorganized and re-equipped. The expected losses of tanks would be unacceptably high and could no longer be compensated for by the production planned for 1943. He emphasized that the tanks produced in 1943 were intended for the Western Front, where Allied landings were expected in 1944. Furthermore, the Panthers still had so many teething problems that it could not be assumed that they would be available in sufficient numbers. Speer placed himself squarely behind Guderian, but the other participants supported Zeitzler's plan, albeit not wholeheartedly, so Operation Zitadelle continued. It would end by undermining the power of German armour.

In the aftermath of the conference, Guderian was approached by von Kluge with the question why he was so unfriendly to him. Guderian said that he had every reason to be, given the events of December 1941, and the two men parted company without patching things up. Sometime later, Schmidt approached Guderian with a letter from von Kluge to Hitler, asking Hitler's permission to challenge Guderian to a duel. Von Kluge knew very well that Hitler had forbidden duels, and Schmidt said that Guderian and von Kluge should settle the matter amicably. Guderian then wrote a letter to von Kluge in which he apologized for his behaviour but also indicated that he felt von Kluge had done him an injustice. The relationship between the two men would remain tense in the next years.

Panther troubles

On 10 May, during a meeting in the Chancellery on Panther production, Guderian seized his chance once again to attempt to dissuade Hitler from launching Zitadelle. Hitler replied that he, like Guderian, was dreading the coming offensive, but that it had to be carried out for political reasons.

In the coming period, the teething problems and production of the Panther would be central to Guderian's agenda. To give an idea of his activities, the table below shows what he was doing in May and June 1943.

Guderian spent a lot of time getting the Panther ready for battle. After its teething problems were fixed, this tank became one of the most valued of the Second World War. This is a Panther Ausf. D of the 12th SS Panzer Division Hitlerjugend, Normandy, summer 1944.

Date	Activity
10 May	Discussion with Hitler in the Chancellery about Panther production.
13 May	Discussion with Speer, and later with Hitler, about Panther production.
14–15 May	Visit to Panzer Abteilung 654 in Bruck an der Leitha, equipped with Ferdinand tanks, also called 'Porsche Tigers'. Visit to the Nibelungen factories in Linz, where Panther tanks and anti-aircraft guns were produced.
26 May	By plane to Paris to visit training for battalion commanders.
27 May	Visit to Panzer Abteilung 216 in Amiens.
28 May	Visit to the training of company commanders in Versailles. Visit to the commanders of the 4th and 16th Panzer Division in Nantes.
29 May	Visit to Saint-Nazaire to inspect the fortifications of the Atlantic Wall. According to Guderian, what he saw did not match the propaganda.
30 May	Return flight to Berlin.
31 May	By plane to Innsbruck for a meeting with Speer.
1 June	By plane to Grafenwöhr to visit Panther Abteilungen 51 and 52, and then on to Berlin.
15 June	Discussion of the technical problems of the Panther (tracks, drive belt and optical equipment).
16 June	Discussion with Hitler in Berlin on delalying deployment of the Panther for the time being due to its technical problems.
17 June	Discussion in Munich with Rommel about the lessons learned in North Africa. In the evening, return by plane to Berlin.
18 June	Inspection of artillery weapons in Jüterborg, then by plane to Berchtesgaden for a meeting with Hitler. Short stopover in Grafenwöhr to hear from the commanders and men of Panther Abteilung 51 and 52 about the the latest news on technical problems.

Despite all the objections from Guderian, Zitadelle was launched on 5 July. The twenty-one divisions that the Germans deployed had a total of 1,715 tanks, as well as 147 StuG IIIs, which increasingly became a permanent feature of the panzer regiments. Although this was a major achievement, the divisions, without exception, had less combat power than in 1941 and 1942.

Guderian visited the front between 10 and 15 July to discuss their experiences with tank commanders, first in the south and later in the north. He gained a good picture of the course of the battle and learned much about the lack of experience of the combat units and the weaknesses of their equipment. He was particularly interested in the Panther and Ferdinand tanks. The ninety Ferdinand tanks that Model had deployed in the north

A nice shot of one of the Tiger tanks of Schwere Panzer Abteilung 505 on the eve of the Battle of Kursk. This Tiger is one of the early series, as can be seen from the commander's hatch on the cupola. To open and close this hatch, the commander had to lean out, an often risky manoeuvre. Later versions had hatches that could be opened and closed from the inside. On the side, barbed wire can be seen; this was to prevent enemy infantry and engineers from climbing onto the tank to place magnetic mines.

proved to be unable to carry enough ammunition for combat, and were extremely vulnerable at close range due to their lack of machine guns. Once they had broken through they were relatively defenceless against Soviet infantry, which they could only neutralize by using their main gun. By the time they reached the Soviet artillery positions they were on their own, and most attacks fizzled out in this way after about 10km.

The Soviets launched their own offensive on 15 July, aimed at Orel. The city was evacuated on 4 August, the same day that Belgorod, the starting point of the German offensive, fell. The failure of Zitadelle was one of the decisive events of the Second World War. The German armoured units, which had been rebuilt and equipped with great difficulty, emerged from the battle badly damaged. Their losses in tanks and armoured vehicles could not be replaced in the near future, and most units would remain well below strength for a long time. The initiative would remain with the Soviets, and from now on the Germans would experience no respite on the Eastern Front.

How critical the tank strength of German tank divisions became in 1943 is shown by the figures below. Before Operation Zitadelle, the average tank strength of the divisions of von Manstein's Heeresgruppe Süd was ninety-five, of which seventy-eight were actually operational. The nominal tank strength of a German tank division, meanwhile, was supposed to be around 150. The failure of Operation Zitadelle heralded a definitive decline in the Panzerwaffe. After almost six months of continuous fighting in the first half of 1943, the strength of the tank divisions had fallen to an average of eighty tanks, of which sometimes only twenty were operational: the repair and maintenance system was unable to maintain combat strength, and most of the tanks were at field workshops or were being literally dragged along when troops moved. One division had only six (!) deployable tanks.

Tiger technology

The Tiger tank was one of the most formidable weapons of the Second World War and was certainly capable of determining the course of battles. However, it had a number of weaknesses. For example, its fuel consumption was very high: the tank contained 540 litres, which could deliver 195km on the road or 110km off it. In comparison, a T-34/76 could, in theory at least, cover 455km on one tank with 480 litres of diesel. Guidelines stated that, ideally, the Tigers would be refuelled within their own lines once they had driven from the assembly area to the front. In order to spare the Tiger's drive train as much as possible, it was advised that the tank be given plenty of room, so that it would have to change gear as little as possible. It was also pointed out that moving these tanks generally led to traffic jams. In fact, the advice was to move them as little as possible, and speed was discouraged altogether because of the great wear and tear that this could cause. The recommended average speed was 10kph during the day and 7kph at night. After heavy fighting, the advice was not to deploy the units for two to three weeks, in order that they could be properly serviced. If this was not done, the number of mechanical failures could multiply rapidly.

The Panzerwaffe was rapidly losing its fighting power. Despite the high quality of the materiel and the professionalism of the men and their officers, the Panzerwaffe's glory had faded, and the great successes of Poland, the western campaign and the early years on the Eastern Front were never again equalled. From now on, the Panzerwaffe could only serve in mounting the strategic defence of Germany, and it would be slowly but surely crushed in the battle with the Red Army.

In the aftermath of Zitadelle, Guderian was struck down by dysentery, caught during his stay in the Soviet Union. He was out of action from mid-July to October. Back at the Inspektorat-General, he discovered that officials from Organization Todt had taken advantage of his absence to redirect production of PzKpfw IV tanks to the production of assault guns. The idea was that they could use the tank turrets thus freed up for bunkers on the Atlantic Wall and in other fortifications. Guderian immediately reversed these decisions and threw himself into the issue of producing mobile anti-aircraft guns. With enemy air forces growing in strength, such guns were now of great importance. Hitler had already approved the production of twin 37mm anti-aircraft guns, but for the time being withheld his approval for quadruple 20mm guns on the chassis of a PzKpfw IV. However, the development of new tank models and variants of existing models continued,

The Jagdpanther combined the outstanding characteristics of the Panther tank with the effectiveness of the 88mm gun.

and on 20 October a number of them could be shown to Hitler. These were models of the Tiger II, the Vomag tank hunter, the Jagdpanther, the Jagdtiger, a Tiger tank with a 380mm mortar, a special model of the PzKpfw III and many others. In other words, designers and manufacturing industry had not been idle.

On 22 October, what Guderian had already warned about happened: Allied air forces shifted their targets from factories that produced aircraft to those producing tanks. The first victim of this was the Henschel plant in Kassel, where Allied bombing created a firestorm that killed 6,000 civilians and seriously injured 8,000. In total, 123,800 people lost their homes. Guderian immediately went to Kassel and addressed the workers in the badly damaged central assembly hall. Many workers and their families had been killed or wounded, and Guderian hoped that his speech would be free of cliché and that his sympathy for them would be apparent. The production of Tiger tanks and 88mm guns was disrupted for months. After the Henschel bombing, on 26 November it was the turn of the factories of Alkett, Rheinmetall-Borsig, Wimag and the Deutsche Waffen und Munitionfabriken in Berlin.

The half-tracks also underwent a metamorphosis during this period due to the different superstructures that became available. This SdKfz is equipped with a quadruple Flak.

The development of tanks in 1943 concluded with a decision to use the chassis of the Czech 38 (t) as the basis for a tank destroyer, the Hetzer. This was to become the basic weapon of the anti-tank battalions of the infantry divisions. Guderian had already urged the development of such a weapon in March, and Hitler fully acknowledged his mistake in not having agreed to it then. The result of this delayed decision was that by the end of 1944 only one third of the anti-tank battalions was equipped with this weapon. With the introduction of the Hetzer, the Panzerwaffe's arsenal for the coming years was complete.

During 1943, the Germans were slowly but surely pushed back along the entire Eastern Front. Losses were very high among all units during the winter battles, but the OKH had no plan for how to replace them. Guderian repeatedly stressed the importance of recalling badly battered panzer units from the front in time to reorganize and re-equip them. In practice, however, these units melted away over time, and completely new ones had to take their place.

The Western Front: the Allied landings

While the Germans were losing ground on all fronts in the east, Guderian focused from the beginning of 1944 on the Western Front. For this purpose he had managed to free up with great difficulty ten panzer and Panzergrenadiere divisions in the past months. These divisions were now being trained and partly re-equipped. In addition, Guderian had another surprise for Hitler, the establishment of the famous Panzer Lehr Division. This was led by Fritz Bayerlein and was composed of the various training and demonstration units of the Wehrmacht, known as 'Lehr' units, hence the name Panzer Lehr Division. This division was considered an elite unit and managed to live up to this reputation in the following year. The fact that even training and demonstration units now had to be deployed at the front indicates how urgent the need for men had become.

The armoured units in the West were commanded by an old friend of Guderian, General Freiherr Geyr von Schweppenburg. His official title was 'General der Panzer Truppen West', but he was subordinate to von Rundstedt, the commander of all units in Western Europe. In addition,

Guderian visiting the Ersatz (training) Tiger Abteilung in Paderborn, June 1943.

Field Marshal Rommel, as commander of Army Group B, was responsible for the deployment of these units along the Atlantic coast in the event of an Allied invasion. Unsurprisingly, this complex command structure would cause the Germans problems in June 1944. In February 1944 Guderian visited France to discuss the situation with Geyr von Schweppenburg and von Rundstedt. All agreed that overwhelming Allied superiority would make troop movements very difficult. In fact, they were only possible at night. In their view, sufficient reserve units should be kept in the rear areas to be deployed quickly if it became clear where the invasion would take place. In this context, it was necessary to prepare the French road network and create alternative river crossings in the form of underwater and pontoon bridges.

Back in Berlin, Guderian was surprised to learn that Rommel's plans had actually called for the armoured units to be stationed close to the coast. When Guderian was back in France in April he asked Rommel for the reason. The answer explained that, based on Rommel's experiences in North Africa and given Allied air superiority, it would be impossible to move large numbers of troops once fighting had broken out, even at night. Rommel also referred to his experiences in Italy. Geyr von Schweppenburg had previously raised the subject of strategic reserves with Rommel and indicated that in his opinion they should be organized as mobile, rapidly deployable units in the hinterland. Rommel had rejected this suggestion at the time and even now could not be dissuaded from his plan. He also pointed out that conditions on the Eastern Front, where Guderian had gained experience, were completely different from those in North Africa, Italy and Western Europe. Hitler agreed with Rommel, perhaps not so much on the basis of his analysis, but because, according to Guderian, the Führer remained committed at heart to the static warfare of the First World War and did not believe in a war of movement. It therefore should come as no surprise that Hitler did not agree with von Rundstedt's and Guderian's proposal as to how the panzer units should be positioned. He gave as his reason that Rommel had more recent combat experience, and Guderian had no choice but to accept this decision.

Ultimately, the Germans had ten panzer and Panzergrenadiere divisions at their disposal in France. They were divided among the different commanders as follows:

- Four under Rommel
- Three under the command of the OKW reserve, including the Panzer Lehr Division
- Three under von Rundstedt in the south to counter a possible landing there.

In addition, two divisions were 'lent' to the Eastern Front, namely the Leibstandarte SS Adolf Hitler and the 2nd Panzer Division; these would return to the west as quickly as possible in the event of an Allied landing. This meant that the panzer units were divided in such a way that concentrated deployment was impossible – precisely what was required on 6 June and the days that followed.

At the time of the landings Rommel was in Germany with his family. Hitler had gone to bed late as usual and could not be woken up. Jodl, responsible for operations when Hitler was not present, was in doubt whether or not to deploy the OKW reserve, since he believed the landings could be a diversionary manoeuvre. The 21st Panzer Division, which was excellently positioned to tackle the British airborne troops, had received orders to postpone a possible counter-attack until it received an order from Rommel himself. In short, in the first hours and days after the landings, the panzer units were partly paralysed by ambiguities in the communication and command structure. In the weeks that followed, they went on to suffer enormous losses, partly caused by errors at the operational level. Thus, the Panzer Lehr Division, for instance, was ordered to advance during the day despite Allied air superiority, and frontal counter-attacks were ordered in areas covered by Allied naval guns.

On 16 June, almost two weeks after the landings, four divisions were still far from the battlefield: the 116th Panzergrenadiere Division was between Dieppe and Abbeville, the 11th Panzer Division was at Bordeaux, the 9th Panzer Division was at Avignon and the SS Division Das Reich was engaged in guerrilla fighting in the south-west of France. In addition, seven infantry divisions were waiting in the coastal areas north of the Seine for possible landings there that never materialized. The result was that a force that could have frustrated a possible landing was not concentrated, was misdirected and was therefore prematurely eliminated. Despite this, in encounters between the two sides the German armoured divisions inflicted terrifying losses on the Allied forces: the Allies repeatedly lost many times more tanks than the Germans. The quality of the German tanks, their crews and their commanders proved time and again to be superior to that of their opponents. But the difficulties encountered by the Germans soon resulted in changes of command: von Rundstedt and Geyr von Schweppenburg had to leave on 29 June, and von Kluge, 'of all people', according to Guderian, replaced von Rundstedt. On 11 July Caen fell and by 17 July, the day that Rommel was seriously wounded when his staff car was strafed by a British fighter plane, the front line ran from the mouth of the Orne via Caen, Caumont and St Lô to the coast at Lessay.

With the loss of Rommel, the fate of the armoured units lay in the hands of von Kluge, a man who, in Guderian's view, had no idea what mobile warfare entailed and was mainly good at small-scale, tactical operations. In the meantime, the Allied units within this bridgehead were preparing for an offensive deeper into France.

The Eastern Front: the collapse of Heeresgruppe Mitte

In the New Year Guderian did not neglect the Eastern Front. During a tête-à-tête breakfast with Hitler in January, Guderian had already raised the subject of field fortifications and defensive zones, in particular in the rear of Heeresgruppe Mitte. In his view, the old German and Russian border fortifications would provide an excellent basis for a defensive line behind the front, which was now moving rapidly westward. This proposal was at odds with Hitler's idea of *Feste Plätze* – cities that would serve as fortresses if the Soviets broke through and would act as a thorn in the side of the Soviet flanks. Stung into action, Hitler leapt to his feet and declared that he, the architect of the West Wall and the Atlantic Wall, knew everything there was to know about building fortifications. Of course he had considered building such fortifications in the East, but the manpower, materials and transport facilities were lacking. Moreover, in his eyes, such fortifications would only tempt commanders to withdraw instead of fighting. Guderian's argument that the transport problems only began east of Brest-Litovsk, and that there were more than enough local materials and labour available, fell on deaf ears. The atmosphere did not improve when Guderian put forward another suggestion: the appointment of a genuine general as Chief of Staff of the OKW to lead the Luftwaffe, Kriegsmarine, OKH and Waffen SS. But Hitler refused to dispense with Keitel, and immediately realized that this was an attempt to limit his own power to some extent.

On 22 June the Soviets launched the long-awaited offensive against Heeresgruppe Mitte, which at that time was under the command of Field Marshal Busch. The offensive was a resounding success, and the Soviet advance westward appeared unstoppable. Over the next few weeks, Heeresgruppe Nord in the north and Army Group A in the south also came under attack and increasingly ceded ground to the Soviets. Hitler's *Feste Plätze* did not work: the Soviets ignored these cities and left the clearing of pockets of resistance to units operating behind the vanguard. The offensive was eventually aimed at Riga in the north and the Vistula near Warsaw in the middle. In total, no fewer than twenty-five German divisions were defeated in a period of a few weeks. This finally broke the fighting power of the German Army. Everywhere, units were scrabbled together and sent to the front, where Busch had been replaced by Model, who was now responsible for both Army Group A and the vacuum left by Heeresgruppe Mitte. Colonel General Harpe would finally take over command of Army Group A shortly afterwards. Under the leadership of Model and Harpe the front was stabilized, and it was possible to start thinking about the defence of Germany itself. But the danger of a breakthrough by the Soviets into East Prussia remained, as a result of their rapid advance and the lack of sufficient German reserves. As the German Army slowly but surely descended into crisis, an event occurred that would once again drastically redirect Guderian's career.

Chapter 11

Chief of the General Staff

The attempted assassination of Hitler

In early July 1944, the situation in the East began to be critical. In order to strengthen the front in East Prussia, Guderian decided on 17 July as an emergency solution to send all deployable units from the panzer training schools in Wünsdorp and Krampnitz to the front in the area around Lötzen, a decision that must have cost Guderian a great deal, since it undermined the training of panzer troops.

The next day, Guderian was approached by von Barsewisch, his old Luftwaffe liaison officer from the Soviet Union, who, representing the conspirators against Hitler, sounded him out. He informed Guderian of the plan to assassinate the Führer but did not indicate when it would happen, since the precise time had not yet been determined. Guderian did not mention this meeting in his memoirs. Instead, he said that von Barsewisch informed him of Field Marshal von Kluge's attempts to contact the Allies with a view to an armistice, that he was surprised by this being done without Hitler's knowledge and that he spent the next few days trying to decide how to deal with it. Accusing von Kluge without further evidence could, after all, be seen by Hitler as Guderian's ultimate attempt to discredit his old adversary. In this way, Guderian cleverly avoided the question of whether and to what extent he was aware of the conspiracy.

On 19 July, Guderian was called by Thomale with the question whether it was possible to postpone the transfer of the training units to East Prussia for a few days, so that on 20 July they could still participate, together with reserve units, in Operation Valkyrie. This was the name given to exercises aimed at countering enemy airborne landings near Berlin or tackling internal unrest. Thomale emphasized that the situation in East Prussia was not so acute that postponement was impossible, and Guderian finally agreed, then

War at the top

On 20 July 1944, an assassination attempt on Hitler, organized by the resistance group around Colonel von Stauffenberg, failed. The bomb placed in the conference room of the Wolfsschanze (wolf's lair) only caused minor injuries to the Führer, and an accompanying coup d'état in Berlin did not materialize. The attack on Hitler on 20 July would change his relationship with the Wehrmacht for ever. He would never trust his generals again, and to show his contempt for them he banned the military salute on 23 July. Suspicion and fear would now characterize the relationship between Hitler and the officer corps. General Ludwig Beck, former Chief of Staff of the Wehrmacht, and General Erich Höpner were compromised by the bomb attack. Beck, the intended head of state to replace Hitler, was given the opportunity to commit suicide, while Höpner was put before a firing squad. There is no doubt that many generals were aware of the plot, but they were faced with a dilemma: if they sided with the plotters, they would be betraying the neutral stance of the army; on the other hand, they did not want to betray their fellow officers.

Another victim was Zeitzler, chief of staff of the General Staff since 1942. His initially good relations with Hitler had deteriorated in the period leading up to the bomb plot, and having failed to make Hitler's perception of events correspond to reality, he had repeatedly asked the Führer to relieve him of his post. Hitler dismissed him the day after the attack and forbade him to wear military uniform. In his place, Hitler surprisingly appointed Guderian, one of the few generals he still trusted. He considered it fortunate that he had placed the Panzerwaffe under its own inspector general: no unit of the Panzerwaffe had had any connection with the conspirators. In addition, the armoured units were too spread out across the various fronts to be able to play an active role in the coup. Hitler, however, immediately made it clear to Guderian that only he could make decisions, and that Guderian should more or less act as his pawn: absolute control over Germany's armies was now in Hitler's hands. In the months that followed, many generals who had served Germany loyally in recent years would be dismissed in short order.

spent the rest of 19 and all of 20 July visiting anti-tank and reserve units at Allenstein, Thorn and Hohensalza. He spent the evening at home, and a motor trooper told him while he was out on a walk that he should expect a call from the high command.

Back home, he heard about the assassination attempt on Hitler, and around midnight he got Thomale on the phone who told him that Hitler wanted him (Guderian) to replace Zeitzler. A plane would pick him up the next morning at 8am at Hohensalza airport to fly him to Lötzen.

It remains unclear to what extent Guderian was aware of the plot to assassinate Hitler. He always denied this in every possible way, but he undoubtedly knew that resistance within the Wehrmacht had slowly but surely become so great that an attempted coup was not an impossibility. The conversation with von Barsewisch was also unambiguous. The training and reserve units that he released on 19 July were supposed to play a crucial role in this coup by occupying the political and military centre in Berlin. Some writers assume that it was not entirely coincidental that Guderian was at home on 20 July. In their view, he was waiting there to see how events unfolded: if the coup had succeeded, he would have sided with the plotters; if it failed, he could declare that he had clean hands.

However, Guderian soon became a suspect. One of the reserve units had received mobilization orders from Fromm, the commander-in-chief of the Ersatz Heer and one of the coup plotters, and had advanced to Berlin in accordance with the orders given to them earlier. Given the unclear situation, they refused to comply with the orders of Major Remer, the local commander who was busy suppressing the uprising, and initially declared – quite correctly – that they would only accept orders from Guderian in his position as Inspektor-General der Panzertruppen.

The scale of the coup attempt should not be underestimated. One of the central figures was Fromm, commander of the Ersatz Heer, alongside General von Stülpnagel, the military commander of France. Ultimately, it also indirectly involved Rommel, who would be forced to commit suicide.

In Lötzen, Guderian was briefed by Thomale and then discussed the situation with Keitel, Jodl and Burgdorf, Hitler's new adjutant since Schmidt had been killed in the attack. Guderian's responsibilities were limited to the Eastern Front along the lines of the OKH's powers. In addition to being

Chief of the General Staff, Guderian remained Inspektor-General der Panzertruppen, but responsibility for day-to-day operations was transferred to Thomale. The post of Chief of the General Staff was only supposed to be temporary; Buhle, who had been wounded by the bomb, would take over after his recovery. Guderian was given the task of replacing the entire OKH, because some had been wounded in the attack and others were suspected of participation in or knowledge of the conspiracy.

Guderian spent the rest of the morning working on new appointments, which he discussed with Hitler in the afternoon. Hitler appeared scarred, bleeding from one ear and with his right arm in a sling. The proposal to replace von Kluge, who according to Hitler had also had prior knowledge of the conspiracy, met with resistance from Keitel, Jodl and Burgdorf. They called him the 'best horse in the stable', while Guderian maintained that von Kluge was not capable of leading large formations. Guderian's attempt to remove von Kluge in this way failed and he decided to let the matter rest. Given his new position and the unique circumstances, Guderian decided to establish a bodyguard for himself, staffed by a number of panzer troops he had known for some time.

Meanwhile, the OKH was in chaos. Not only were posts not filled, but Zeitzler had decided that the OKH would be moved to Zossen near Berlin; a large number of personnel, including the communications network, had already been moved there. Guderian decided to reverse this decision because he wanted the OKH to be positioned close to the OKW in East Prussia. Guderian then filled the vacant positions with officers whom he knew well and with whom he had worked closely over the past years. It took some time for the members of this new team to arrive from all points of the compass, and even longer for the OKH to become a well-oiled machine again.

However, Guderian would not be able to escape the coup attempt and its aftermath. Hitler had determined that the soldiers involved should not be tried by a military court, but by a so-called Volksgericht, a court that applied special laws and gave verdicts that could not be appealed against. For this to happen, the soldiers concerned first had to be discharged from military service after investigation by a special court, chaired by von Rundstedt and with five members including Guderian. Guderian tried to get out of this task by stating that the dual role he now filled, Inspektor-General and chief of staff of OKH, took up too much time. This request was rejected, but he was permitted occasionally to send a substitute. After Keitel insisted, Guderian did finally attend a few sessions. The members of the court found themselves in an almost impossible position: they had to pass judgment on their own comrades, colleagues whom they had known well for years. These officers were very open to the court about their knowledge of the conspiracy and the contacts they had made; thus a complete picture of the scale of the conspiracy was rapidly established. In addition, the possibilitiy of a coup had been so widely discussed in recent years that lists of ministerial posts were more or less openly circulated; few men could deny having had any knowledge of this. Guderian and his fellow officers had no choice but to struggle through the court sessions and protect their colleagues from themselves as much as possible.

The military situation

The military situation that Guderian had inherited from his predecessor was extremely worrying. In the West, the Allies had managed to break out of their bridgehead and encircle the divisions that Hitler had deployed for a counter-attack. Most of these fifty divisions had been destroyed in the aftermath of the Normandy landings, and von Kluge, suspected by Hitler of approaching the Allies for an armistice, had committed suicide when he received the message to report to headquarters. He had been replaced by Model, the commander who had been able to restore some stability to the Eastern Front after the collapse of Army Gruppe Mitte. The situation in the West was different, however, and in the first weeks of Model's command the German units had no choice but to withdraw in an orderly manner towards Belgium and the Netherlands in August and September. The problem was that there were no fortifications or reinforced lines on which the Germans could fall back; only the West Wall provided any protection. However, the West Wall was no longer a genuine fortification: most of its artillery had been moved to the Atlantic Wall, where it had been lost during the Allied landings. The Allies then pursued the Germans with such determination that due to the lack of sufficient personnel, the fortifications – if any – had to be rapidly abandoned.

With his chief of staff, Generalleutnant Wenck, Guderian stands bent over the map table in the OKH.

Meanwhile, on the Eastern Front, the Soviet offensive had run its course by the end of June. The Red Army had manoeuvred itself into an excellent starting position to advance on Berlin via Poland and Prussia. But Stalin had decided to exert more pressure on the vulnerable and politically desirable Balkans first. In August and September, the Soviets launched offensives that delivered major territorial gains in Romania, Bulgaria and Hungary. The Romanian and Bulgarian governments felt the time was ripe to switch sides, but they would not do so without suffering. The consequences for Germany were very great: the loss of Romania also meant the loss of its oilfields. Since the German synthetic gasoline plants were far from being able to meet demand, possession of the Hungarian oilfields near Lake Balaton now took on special significance.

New equipment

In the meantime, tank production was experiencing unprecedented growth, partly due to the investments made in 1943 in new production facilities. The second generation of tank hunters (Jagdpanzer) had made its debut in late 1943 in the form of the Hetzer, Jagdpanther and Jagdtiger.

The Hetzer, based on the chassis of the obsolete PzKpfw 38(t), was distributed to the anti-tank battalions of infantry divisions. Its superstructure had angled corners, and it had 60mm armour and a potent 75mm L/48 gun on the right. On top was a machine gun that could be remotely controlled from the compartment and could rotate through 360°. The Hetzer was admired for its design, but space for the four crew members was limited. The commander sat to the right of the gun, the driver, gunner and loader sat to the left. The Hetzer could carry forty-one shells and, despite its limited space, proved to perform excellently at the front.

The Jagdpanther was in some respects the opposite of the Hetzer. With its 45 tons and 88mm L/71, it could easily take on any tank that the Allies or the Soviets could throw into battle. Probably the best tank hunter of the war, it was built on the chassis of a Panther tank and had a sloping superstructure with 80mm armour in front that provided protection equivalent to 160mm of vertical armour. The Jagdpanther's drive train was very reliable, its excellent 700hp Maybach engine gave it a speed of 45kph, it carried eighty shells, and the five-man crew could defend themselves against attacks at close range with a machine gun in the sloping front section. The tank's limitations were the manoeuvring space for the gun and its height of 2.72m. A total of 228 tanks of this type were produced in 1944.

The Jagdtiger can be seen as the finest of all tank hunters. Built on the chassis of the Tiger, its 128mm L/55 gun made it the most heavily armed tank hunter of the Second World War. The shells were split into two parts to make them easier to handle, which made its rate of fire lower than that of other tank hunters. It had 250mm frontal armour, with 80mm on the sides and 40mm on top. Despite its weight of 72 tons, it could reach a very acceptable speed of 37kph. The limitations of the Jagdtiger lay in the technical complexity (and sensitivity) of its mechanical parts and in the fact that its weight made it difficult to tow. A total of forty-eight tanks of this type were produced in 1944.

All these new tank types were mainly deployed on the Western Front. On the Eastern Front they were only available in limited numbers and, like the sPzAbts, were deployed as a 'fire brigade' in times of crisis.

Tank production reached its absolute peak in the summer of 1944, and a number of formidable weapons were added to the Panzerwaffe's arsenal, including the Tiger II (called 'King Tiger' by the Allies) and the Jagdpanzer IV/L70. In total, more than 19,000 armoured vehicles rolled off the production lines in Germany and other European countries in 1944, despite Allied bombing. With these weapons, Hitler was convinced that in the winter, when Allied air forces would be hampered by the weather, an offensive in the West would be possible again.

German production numbers of main types of armoured vehicles 1943–1944.

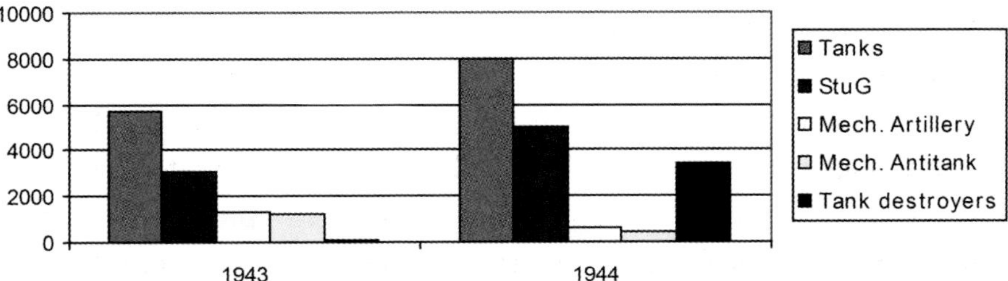

Production of armoured mechanized vehicles during the war years.

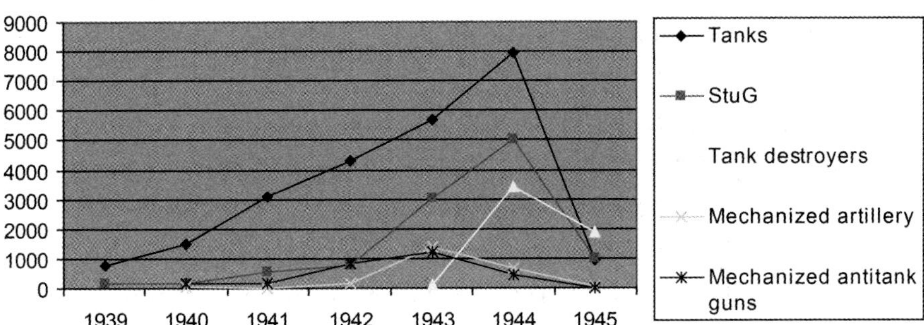

In total, this represented 8,328 medium and heavy tanks, 5,751 assault guns, 1,617 mechanized anti-tank guns, 1,246 mechanized artillery pieces on various chassis and many other variants.

Distribution of the main types in 1943 and 1944

Sort	Type	Numbers	
		1943	1944
Tanks	PzKpfw	3,013	3,126
	Panther	1,786	3,749
	Tiger I	647	623
	Tiger II	-	377
Assault guns	StuG III	3,011	3,840
	StuG IV	31	1,006
	StuG M42	159	135
	StuG M43	-	18
Mechanized artillery	PzKpfw 38(t) with sIG33	215	154
	PzKpfw II Wespe	514	162
	PzKpfw IV Hummel	368	289
	Superstructures on various chassis	256	52
Mechanized anti-tank guns	PzKpfw II 75mm Marder II	204	-
	PzKpfw 38(t) 75mm Marder III	618	323
	PzKpfw III/IV 88mm Nashorn	345	133
Tankhunters	Ferdinand	90	-
	Hetzer	-	1,457
	Jagdpanzer IV	-	769
	Pz IV/70	-	767
	Jagdpanther	-	226
	Jagdtiger	-	61

Despite all the efforts of Speer and the German armaments industry, the attrition of weapons and vehicles was so great that Germany was slowly but surely losing the battle in this area too.

Courland and the defence of Germany

Among the many problems that Germany had to deal with at the end of the summer of 1944, three particularly concerned Guderian. One was the danger of the encirclement of Heeresgruppe Nord in Estonia (Courland); the second concerned defences in Germany itself and the supply of weapons to local civil defence units (*Volkssturm*); the third was the mobilization of new reserves for the army. The fate of Heeresgruppe Nord was inextricably linked to his fear that the entire Eastern Front would collapse.

Hitler insisted, however much Guderian and others tried to change his mind, that Courland not be abandoned. The Führer's line was that no territory should be surrendered, something he had undoubtedly been right about in the past when the German forces were still more or less on a par with the Soviets. He also stated that the German units in Courland tied up a large number of Soviets and were therefore serving a very useful purpose. The disagreement over Courland continued as long as Guderian was chief of staff, and as a consequence, Heeresgruppe Nord would go down fighting in the coming months without having made a significant contribution to the defence of the homeland. No fewer than nineteen infantry and five panzer divisions were lost with it.

Fortifications

In the second half of 1944 Guderian began planning fortifications on the Eastern Front. Together with Lieutenant Colonel Thilo he drew up a plan which he sent to the relevant authorities before it had received Hitler's approval. Only then did he submit the plan to Hitler, who was 'not amused', a clear signal to Guderian not to try this tactic again. The fortifications were mainly built by volunteers, women, the very young and the old, the only untapped labour force in Germany. Guderian's original plan was to man the fortifications with units built around men who were no longer able to serve at the front but who would be perfectly capable of defence under good leadership. First, 100 fortress infantry battalions would be formed, supplemented by 100 artillery batteries. Machine-gun, engineering and communications units would soon follow. The idea was that the fortifications would be provided with three months' worth of supplies. Communication centres were installed and fuel stocks were built up. Guderian visited many of the fortifications himself to check whether they met expectations.

Guderian got help from all sections of the Wehrmacht, and many Gauleiters (district party leaders) also did everything they could to support him. But Guderian found his plans were thwarted: before the first units were fully trained, 80 per cent of them were sent to the front in the West, and he was only informed of this after the transfers had already been made. Protest was fruitless, and when the front collapsed in the West, these units were crushed. They had made no significant contribution to the battle.

The same story applied to armaments. Guderian wanted to use the many thousands of captured weapons, now stored in various arsenals, to arm the defence units. Jodl and Keitel denied that there were such arsenals, but General Buhle, the Chief of the Army Staff of the OKW, declared that there were indeed stocks of these weapons, and that they had been properly maintained and greased monthly. Guderian ordered that they be placed at the most crucial points in the various fortifications and that the personnel involved be trained in their use. However, Jodl managed to intercept this order and ordered that every artillery piece larger than 50mm or with more than fifty shells be sent to the front in the West. This meant that Guderian no longer had the only guns capable of stopping Soviet tanks; the remaining 37mm and 50mm guns were no match for the T-34s or other heavier Soviet tanks.

Many fortifications and strongholds in the East ultimately remained largely unmanned and were therefore unable to support the retreating German units. In practice, some of the fortifications that were operational ultimately proved to be excellent, such as those at Königsberg, Danzig and Breslau.

Landsturm

After it proved impossible to man the fortifications, Guderian's next idea was to establish a *Landsturm* (militia) in the eastern provinces to defend the homeland. These units would only be mobilized if the Soviets broke through the front. They would be commanded by regular officers and consist of men who were capable of active service but who had been released from it for various reasons, for example their essential jobs in the armaments industry. Guderian presented this proposal to Hitler and suggested that the SA should be given the task of setting up these units. He had already contacted Schepmann, the SA Chief of Staff, about this, in Guderian's eyes a sensible man who was well-disposed towards the army. Hitler agreed to the plan but felt that the units should be the responsibility of Reichsleiter (party secretary) Bormann. He also wanted them to be called '*Volkssturm*'.

Bormann initially took no action, but after repeated urging from Guderian, he instructed the Gauleiters throughout Germany to establish *Volkssturm* branches. As a result, the *Volkssturm* became unmanageably large and had neither sufficient trained staff nor enough weapons. Moreover, party officials were more interested in the political reliability of the men than in their military qualities. As a result, members of the *Volkssturm*, who were prepared to give their all, were mostly trained in giving the correct Hitler salute rather than in the handling of weapons.

Volksgrenadiere divisions

Hitler also had his own ideas about the formation of new fighting units. In September he ordered the establishment of so-called Volksgrenadiere Divisions, because it appeared to him that the army was unable to hold positions and territory. Himmler himself, commander of the Ersatz Heer since the attack of 20 July, was busy creating these divisions, built around politically reliable men and officers. It was an idea that he and Hitler had long dreamed of. The units were also staffed with a German version of the political commissars of the Red Army, thus installing *Fremdkörper* (foreign bodies) in the military, men who were primarily accountable to the party not the army. When reports about the functioning and political reliability of various units and their commanders began to reach Hitler via party channels, Guderian tried as quickly as possible to put an end to the further politicization of these Volksgrenadiere divisions. These divisions themselves were created at a rapid pace. Only twelve were completely new; the remaining thirteen were formed from units that had been reduced in battle to such an extent that they could not be rebuilt. The men of these Volksgrenadiere divisions had completed an average of just two months of training, and their combat effectiveness varied from 'as good as that of an old division' to that of an 'armed mob' – not surprisingly, considering the lack of training and team building and the shortages of weapons and equipment.

The character of the armoured divisions also changed dramatically during this period. The Germans had accepted the fact that they were no longer able to field the ideal division of 1941, fully equipped with tanks. The structure of armoured divisions now was as follows. One of the regiments was called 'motorized infantry', although only one infantry battalion in this regiment could use half-tracks – at least in theory; in practice often only one company had them. This was compensated for by more and heavier weapons. The tank regiment had only two battalions, one with PzKpfw IVs and one with PzKpfw Vs (Panthers). One battalion of the artillery regiment had eighteen mechanized guns.

Guderian and Hitler constantly argued about the best way forward. The attack on his life had had a devastating effect on Hitler's personality. From then on he assumed that everyone was lying to him, and he no longer trusted anyone. He regularly lost his temper, his language became coarser, and his behaviour worsened from month to month. While in his inner circle circle the affable and civilized adjutant Schmidt had in the past had a tempering effect on Hitler, now all restraints fell away. Unsurprisingly, the atmosphere within the high command was febrile to say the least: if we are to believe Guderian's memoirs, there was almost constant quarrelling from September 1944 onwards. Hitler often became uncontrollably furious when he heard that territory had been given up, and the suggestions of Guderian and others were usually rejected. The Führer constantly accused Guderian that he and the rest of the officer corps did not have sufficient strategic insight into political and economic matters to understand what was happening. In addition, the Führer and the party leadership assumed that the military was constantly feeding them incorrect information. Hitler considered news about the Soviet units on the Eastern Front, for instance, to be particularly unreliable.

Meanwhile, by mid-December, all panzer and panzergrenadier divisions on the Eastern Front had gradually been withdrawn from the front line. They were re-formed into four groups to become a mobile reserve behind the front. It was not possible to regroup infantry divisions behind the front in the same way, because there was a chronic shortage of men. It was only possible to station one infantry division as a reserve near Kraków.

In addition to differences of opinion about fortifications and other issues, Hitler and Guderian also differed in their views about the position of the so-called *Hauptkampflinie* and the *Grosskampflinie* in the east. The former was the usual front line, to be defended under normal circumstances. The latter would serve as a defence line in the event of a large-scale Soviet attack in one of the sectors. The officers at the front wanted to build this

Visiting the Hitler Youth, who were ordered to help build positions in the East.

Grosskampflinie 18–20km behind the *Hauptkampflinie*, carefully camouflaging it and manning it with garrison units. They also wanted a standard order that the units should be allowed to withdraw on this line as soon as a Soviet artillery offensive broke out; in the area of the *Hauptkampflinie*, only outposts would remain. The Soviet barrage would thus be a waste of effort, their carefully planned attack would be fruitless and they would become stuck on the *Grosskampflinie*. When Guderian suggested this to Hitler, the Führer reacted furiously. Based on his experiences in the First World War, he demanded that the *Grosskampflinie* be built no more than 2–3km behind the *Hauptkampflinie*. This decision would cost the Germans dearly in January, when their *Hauptkampflinie*, *Grosskampflinie* and reserves were buried under the flood tide of the Soviet offensive. When this happened, Hitler's anger was directed at those who had built the *Grosskampflinie* so close to the *Hauptkampflinie*. Guderian reminded Hitler that he himself had ordered this and suggested that the minutes of the meeting in question be called for. The minutes were duly obtained, and Hitler read them aloud. After a few sentences he stopped, realizing that he was the one responsible for this mistake. But by then it was too late, and the Soviets were able to force a breakthrough.

Hitler's attitude in this case is typical of the entire period. He continued to believe that he had the best tactical insight in the high command because he was the only one with experience at the front. And he had a point: most of his military advisers had much less active front-line experience than he did. Because, moreover, party comrades such as Göring and von Ribbentrop continually fed him the illusion that he was a great military leader, he was unwilling to learn from others. As he often said when reprimanding his generals:

> You don't have to try to teach me anything. I have been in charge of the German Army for the past five years and during this time I have gained more practical experience than the gentlemen of the General Staff could ever hope for. I have studied Clausewitz and Moltke and read all of von Schlieffen's plans. I have a sharper eye on things than you do!

The Ardennes offensive

While these discussions about the Eastern Front were taking place, in the autumn of 1944 Hitler had become completely obsessed with 'his' offensive that was supposed to bring the German units to Antwerp and drive a wedge between the British and Americans. According to Hitler, this would force the Western Allies to the negotiating table. The offensive was planned for the middle of November. After the conclusion of this – limited – offensive, there would be sufficient time to transport the released units to the Eastern Front, assuming that the Soviets would not launch their winter offensive before the New Year, given the relatively mild autumn and the fact that the front in the East had become temporarily static. In the light of these considerations, the Eastern Front now occupied second place in Hitler's mind. In order to bring the necessary units into the field for his offensive, Hitler had set his sights on the reserves that the army had managed to mobilize with great difficulty in the preceding months and that were primarily intended for the Eastern Front. These two Panzer armies were used for the Ardennes offensive instead: the Fifth under General von Manteuffel and the Sixth under SS Colonel General (Oberstgruppenführer) Sepp Dietrich. The main spearhead was on the right, northern flank of the offensive, in the area of the Sixth SS Panzer Army, which contained the best-equipped units of the Waffen SS. The Fifth Panzer Army operated in the centre and on the left, and on the southern wing the Seventh Army under General Brandenberger was deployed with the aim of protecting the left flank of both Panzer armies. Brandenberger, however, did not have the necessary mobile units to fulfil this important task.

The offensive originally had a limited goal in the plans of Model and von Rundstedt: to neutralize the Allied units north of the Meuse in the area between Aachen and Liège. Hitler, however, had a more ambitious idea: the units would advance via Brussels to Antwerp and neutralize the Allied units north of this line. Antwerp would no longer be able to function as a supply port for the Allies, and this would, in Hitler's optimistic opinion, force the Allies to the negotiating table. As described earlier, the offensive was planned for mid-November so that the armoured units involved could be deployed to the east in time to resist the Soviet winter offensive. In reality, the Ardennes offensive had to be postponed until mid-December, thus putting further pressure on the already critical schedule for the deployment of the units involved on the Eastern Front.

On 16 December, the Ardennes offensive was launched, and initially the American units were completely overwhelmed. Units of the Fifth Panzer Army managed to achieve deep penetration of enemy lines, with some reaching the banks of the Meuse. The Sixth SS Panzer Army made less good progress, due to a combination of traffic jams on the many narrow, slippery roads, shortages of fuel and a failure to capitalize quickly enough on the surprise effect of the attack. As the Seventh Army encountered stubborn resistance by the Americans, von Manteuffel was soon forced to deploy his tanks on the threatened left flank, which dashed the hope of a large-scale breakthrough.

By 22 December it was clear that a less ambitious goal had to be settled for and that, in view of the impending Soviet attack in the East, it would be wise to break off the offensive as soon as possible. However, Hitler and the OKW remained completely obsessed, and even after 24 December, when the Allied air forces were once again supreme due to the appearance of cloudless skies, neither Hitler nor the OKW wanted to halt the offensive.

Guderian explains the situation on the Eastern Front at a press conference in 1945.

The Soviet offensive

While six panzer divisions were deployed for the Ardennes offensive, Guderian had only fourteen panzer divisions at his disposal in the East to hold back the Soviets along a 1,200km front. The German front was, in Guderian's words, a house of cards: if it was breached in one place, the whole thing would collapse. He estimated that the Red Army outnumbered the Germans by 11:1 in infantry, 7:1 in tanks, 20:1 in artillery and 20:1 in aircraft. Guderian discussed the latest intelligence with Hitler on 24 December in the presence of Keitel, Jodl, and Burgdorf, among others. Hitler completely rejected the figures and angrily asked who was responsible for them. He was convinced that the Soviet units were largely fakes, like the notorious Potemkin villages, and that they had no intention of attacking in the near future.

That evening during dinner, Himmler confirmed that these were the views of the party leadership, saying to Guderian, 'I (do not) think that the Soviets will attack at all. It is an enormous amount of bluff. Your data is far too exaggerated. Your people are far too alarmed. I am convinced that there is nothing wrong in the East.'

Guderian could do little against such gullibility. Much more dangerous was Jodl's position: he was convinced that the Allies had been thrown off balance by the now stalled Ardennes offensive, and that a number of limited offensives could bring success and ultimately paralyse the enemy. On 1 January, an offensive was therefore launched in the Alsace-Lorraine region with the aim of capturing Saverne and even Strasbourg. Guderian's request to deploy to the east any units that might be released was rejected, because, according to Jodl, 'the initiative that we now have should not be relinquished.' And so the army leadership muddled on.

But things were to get worse still. The Soviets were still advancing in the south of the front and on 24 December they encircled Budapest and reached Lake Balaton. On 25 December Hitler, without informing Guderian, ordered Gille's SS Corps with its two SS Panzer divisions to move from the reserve on the Eastern Front to Budapest, with the aim of relieving the city. Guderian was dismayed, but his protests fell on deaf ears. This move reduced the strategic reserve in the east to twelve panzer divisions. Furthermore, preparations for the coming battles were increasingly being frustrated by the Gauleiters. They, and Hitler, refused to consider evacuation of the civilian population in the most threatened areas and accused the generals of scaremongering and defeatism. The operational zone of German units was allowed to be only 10km deep, and this caused problems, for example, when setting up gun emplacements, which had to be much further behind the front, in areas that fell under the jurisdiction of the Gauleiters. In these areas, by orders of the party, no ditch could be dug and no tree felled without permission from the local authorities.

Guderian expected the Red Army to resume its attack from across the Weichsel (Vistula) into the heart of Germany on 12 January. His assessment proved correct: on that date, a large Soviet force broke out of the Baranov bridgehead in southern Poland and advanced towards Silesia. At the same time, other Soviet units crossed the Narev further north and advanced into East Prussia and Pomerania. On 20 January, the first Soviet units set foot on German soil: the end was now at hand.

In previous days, argument had raged between Jodl and Hitler on the one hand and Guderian on the other. The subject was the deployment of the units in the west, now that both Jodl and Hitler had to admit that the greater threat came from the East. More specifically, it was about the deployment of the 6th SS Panzer Corps, which Hitler had ordered to be transported to Hungary. On 15 January, Hitler had also intervened by moving the Panzer Corps Gross Deutschland, with the Panzer Grenadiere Division Gross Deutschland and the Fallschirmjäger Division Hermann Göring, from the reserve in East Prussia to Kielce, to prevent a possible breakthrough by the Soviets towards Posen. When Guderian learned from Jodl on 16 January the new destination of the 6th SS Panzer Corps, he exploded, at which Jodl merely shrugged. At the subsequent conference, Guderian immediately brought this up, but Hitler indicated that the purpose of the offensive was to protect the small Hungarian oil fields, repeating that his generals did not understand the economic dimensions of war. It was during this conference that Hitler reacted angrily to the fact that the *Hauptkampflinie* and the *Grosskampflinie* had been so close to each other. Since scapegoats had to be found, a carousel of dismissals and new appointments began – not for the first time. The only positive outcome was that the offensive in the West was finally halted, and the German units there went on the defensive.

From the above developments it is clear that Guderian was increasingly losing control over the course of events. Hitler simply did what he thought was best, without paying much attention to Guderian or others. On 25 January Guderian asked von Ribbentrop to join him in urging Hitler to agree to an armistice. Von Ribbentrop accepted Guderian's suggestion and put out feelers to the Allies. On 25 March Guderian also tried to involve Himmler in this initiative, but he rejected it. Himmler, however, then sent out his own representatives to attempt to negotiate an armistice. Later in the same month, Guderian and Speer tried to prevent Hitler from destroying crucial parts of German industry and infrastructure. Speer and Guderian believed that it was essential that Germany retained a chance to recover after the war. Together with Jodl, he also managed to prevent Hitler from renouncing the Geneva Convention. In the threatening atmosphere of a rapidly approaching end, it became clear that Guderian's days in the army leadership were numbered and that he was physically and mentally at the end of his tether.

The end for Guderian

Guderian had little time for staff work, since Hitler wanted him to attend the daily meetings in the Führerbunker, which increasingly took up most of each day. Guderian estimated that, including travel back and forth between Zossen and the Führerbunker, this took him eight or nine hours a day. He had almost no opportunity to visit his units at the front, especially since the Allied air forces increasingly made travel impossible. In addition, the series of defeats constantly undermined his position.

It was therefore no surprise when Hitler asked Guderian to go on sick leave on 21 March 1945. Guderian replied that there was no one at the OKH at that moment who could replace him, since both of his closest colleagues had been wounded shortly before. Hitler left it at that for the time being, but a week later, when Guderian tried to protect one of his field commanders from a fit of the Führer's rage, he had had enough and ordered him to take six weeks' holiday. Keitel, with no idea of the reality of the battlefield, suggested Bad Liebenstein as a place for Guderian to recuperate. Guderian told him that it was already in American hands; when offered the alternative of Bad Sachsa in the Harz, Guderian replied that he would prefer somewhere that was not going to be taken by the enemy in the next forty-eight hours. In fact, Guderian travelled with his wife to Munich, where he was treated for his heart condition in the weeks that followed. When the six weeks were up, he reported back on 1 May to the Inspektion der Kraftfahrtruppen, which had in the meantime been evacuated to Tyrol. On 10 May he was taken prisoner by the Americans. His war had come to an end.

Chapter 12

Guderian: a Retrospective

While in captivity, Guderian was subjected to intensive interrogations, and after charges of crimes against humanity were dropped, he cooperated realistically with the Allies and wrote a series of evaluations for the US Army Historical Division. After his release he moved to a small house in Schwangau, where he wrote his autobiography, *Errinerungen eines Soldaten* (*Memories of a Soldier*), which was published in 1951 and translated into English the following year under the title *Panzer Leader*. The book became an immediate bestseller in the United States and Great Britain and laid the foundation for his reputation as the founder of the German Panzerwaffe. Guderian died on 17 May 1954.

Looking back at Guderian's life, his book *Panzer Leader* appears central to the picture that emerges. With this book Guderian claimed a permanent place in military history and became, together with Rommel, perhaps the most famous German general of the Second World War. That his role in the formation of the Panzerwaffe was in reality less prominent than he would have us believe should have become clear from earlier chapters. Many writers have uncritically adopted Guderian's account, without examining the context in which he operated or realizing that, partly because of his position in the hierarchy, Guderian could never more or less single-handedly have developed the concept of tank warfare, let alone been the founder of Blitzkrieg.

The first thing to remember is that Guderian came from a classic Prussian military environment. Kulm, his birthplace, is in the area between the Oder and the

Guderian in 1950.

Memel, the historic region of Prussian feats of arms. For Germans like Guderian, the Prussian presence in the Baltic States for 500 years was an important historical fact. They believed that they had brought order, civilization and prosperity there for centuries. His involvement with the Eisernen Division after the First World War and his dubious attitude when it was recalled show how important Guderian's ties were with the area. What for us may be a series of episodes without any connection, many of which did not take place on German soil, had a completely different meaning for Guderian and many other Germans.

There is no evidence that Guderian was affected by the loss of the West Prussian provinces after the First World War, or that their reconquest played a significant role in his further career. It does explain, however, why Guderian made no critical comments, even in his memoirs, about the conquest of Poland, which was after all a sovereign state. In his eyes and those of other officers in the Wehrmacht, this was no more than the reversal of an unjust decision taken by the Allies after the First World War. The war in the East was in fact a common thread through Guderian's wartime life, first as a field commander in 1941, later in 1944 as chief of staff, since the OKH was responsible for that front, while the OKW took responsibility for the front in the West.

When we look at his career in tanks, a number of things immediately stand out. First of all, Guderian had no experience with them during the First World War. The fact that he was able to look at issues dispassionately as a staff officer may have contributed to his later analysis and the line he took on the basis of it. Central to this view was the belief that trench warfare, with its appallingly high costs in men and materiel, was not an option for future wars. His posting after the First World War to the Inspektion der Kraftfahrtruppen shows that the hierarchy had not forgotten his attitude in 1919 to the actions of the Eisernen Division. But somehow, with his sharp intellect, Guderian managed to lay the foundation for his later role in the Panzerwaffe in precisely this position, which he rightly regarded as literally far from the front. His analyses of supply and transport problems led him to the conclusion that motorized and armoured units would play a central role in future conflicts. He was certainly not the only person in Germany or abroad who was thinking about this subject, but he managed to give his ideas ever clearer shape in the course of the 1920s. The fact that he had not seen the inside of a tank until the late 1920s makes this all the more remarkable.

In autumn 1929, Guderian was approached by Oswald Lutz to take command of one of the transport battalions under his command. This request largely determined Guderian's further career. Lutz played a central role in the development of armoured vehicles from 1924 to 1938, first as head of department within the Heeres-Waffenamt, later in 1928 as chief of staff of the Inspektion der Kraftfahrtruppen. The transport battalions of the Kraftfahrtruppen of the seven Wehrkreise were under his command. Improbable though this may have seemed, these transport battalions would grow in the years that followed into the Panzerwaffe, a development in which Lutz and Guderian both played important roles. The dismissal of Lutz in 1938 allowed Guderian to emphasize his own influence in his autobiography and to minimize that of Lutz. In reality, Guderian was one of a number of players involved in the further development of the Panzerwaffe; many of the others would be less well-known, but were equally effective commanders. One of the important ways in which Guderian distinguished himself from them was his analytical and intellectual ability, which ultimately led to publications such as *Achtung-Panzer!*, as well as his autobiography. These books secured Guderian a place among the most famous generals of the Second World War and ensured that his claims about his role in building the Panzerwaffe were seldom questioned.

Of course, his career consisted of more than just theory: in the period 1939–1941 he commanded several armoured units. During the Polish campaign, as well as later in the West and during Operation Barbarossa, his leadership style was characteristic of many German officers: always ready to take personal risks and set a good example by investigating situations at the front himself, thus experiencing battle in all its intensity. All this made him a particularly inspiring leader, despite often coming into conflict with his superiors due to his self-will and sometimes his refusal to obey orders. This could partly be explained by his determination to keep driving his units forwards. The fact that this meant that contact with other units, particularly the infantry, was often lost, was something he accepted. He also had a gift of being able to predict the course of events with uncanny accuracy. All these qualities would develop fully during the campaign in the West and during operations in the Soviet Union. Despite the fact that he would often turn up at unexpected times and places, he never lost his overview of the whole battlefield. He spent a large proportion of his time at the front among his men, sometimes even in front of his advanced staff headquarters.

We should not forget that his experiences as a liaison officer also laid the foundation for the excellent C3I capability of the German Panzerwaffe, which enabled the tanks to maintain control and coordination when they were operating over a large area. His special status within the Wehrmacht was reflected in the large 'G' for Guderian on the vehicles of his units: no other general literally left his mark in this way. His men always showed that they were proud to serve under Guderian.

Guderian's strategic ability is difficult to assess. He was a declared advocate of Moscow as the ultimate goal of Barbarossa. The road to Moscow was open in the autumn of 1941, but it is unclear whether the city's capture would have broken Soviet resistance and whether and to what extent the Germans would have been able to protect the long flanks and secure the supply lines which an advance to Moscow would have necessitated. Moreover, the Soviets seemed to pay little attention to these and other strategic considerations; they simply fought on, supplied with weapons by industry located far to the east.

Regular conflict with his superiors can also be seen as a thread running through Guderian's career. The animosity between von Kluge and Guderian is one example, but it also indicates the divide between a representative of the established order, namely the infantry, and the advocate of an innovative but as yet unproven weapon, the tank. Both were right: on the one hand, tanks served to force fast and deep breakthroughs, while on the other hand, the infantry were then needed to consolidate conquered positions and to protect them against counter-attack. Conflict with his superiors was ultimately the reason why Hitler relieved Guderian of his position as commander of his own Panzer army in December 1941.

Miraculously, Guderian achieved a comeback in 1943 and came close to the centre of power. His role as Inspektor-General should not be underestimated, and his influence on armour was probably more important in that period than in the 1930s. Together with Speer, he brought order to the production process by setting clear priorities, and he shortened the innovation cycle by modifying existing weapons and developing new ones based on the most recent experiences from the front. It is impossible to reconstruct accurately how he managed this within the German military-industrial and political complex. It is clear, given the way that he dealt with Himmler, Goebbels and other representatives of the Nazi party leadership, that he suffered from a certain political naivety. They probably laughed at him behind his back, seeing him as a specialist in the field of armoured warfare but a complete innocent in politics. But it was from his specialized knowledge that he rightly derived his authority, an authority that no one ever doubted.

In the last period of his career, as chief of staff, he could no longer play a major role. After the assassination attempt, Hitler was convinced that no officer could be trusted and that the entire Wehrmacht, including the General Staff, was peopled by traitors. He was therefore prepared to delegate very little power to them. Before the assassination attempt, Hitler had personally directed actions outside the Soviet Union, but after the attempt, the Eastern Front would also fall under his juridiction, so Guderian could only translate the Führer's decisions into orders to the troops. Guderian's efforts were aimed at, among other things, preventing Hitler's anger from continually coming down on the heads of field commanders who were unable to rescue the fortune of the Wehrmacht on the battlefield. Guderian was certainly not a pawn of Hitler; he was an officer who was determined to protect the reputation of the German Army. A long series of quarrels with Hitler, which testified to fundamental differences of opinion, characterized this period, which was concluded unceremoniously with Guderian being forced to take sick leave.

In retrospect, one cannot help but be impressed by Guderian's career: for more than a decade and a half he was at the heart of events that were decisive for the twentieth century. His career was not planned, but rather took the form of a series of appointments largely thanks to his sharp intellect, perseverance and idiosyncratic approach. The diversity of positions he held distinguishes him from the many other well-known generals who, however capable, were never involved in major, far-reaching innovations or important political developments. Guderian can therefore rightly be seen as a driving force behind one of the most important military innovations of the twentieth century – the tank, a weapon that still captures the imagination of many today.

Bibliography

Adair P. (1994), *Hitler's Greatest Defeat; Disaster on the Eastern Front*. London: Rigel publications

Alger, J.I. (1982), *The Quest for Victory*. Westport: Greenwood Press

Army Department (1951), 'Rear area security in Russia'. Washington: Office of the Chief of Military History

Auteurs van Command magazine (1988), *Hitler's Armies; the evolution and structure of German armed Forces, 1993–1945*. New York: Combined Books, Inc.

Broekmeyer, M. (1999), *Stalin, de Russen en hun oorlog*. Amsterdam: Uitgeverij Jan Mets

Carver, M. van (1986), *Dilemmas of the Desert War*. London: B.T. Batsford Ltd

Corum, J.S. (1992), *The Roots of Blitzkrieg; Hans von Seeckt and German Military Reform*. Lawrence: University Press of Kansas

Creveld, M. van (1974), *Fighting Power*. Westport: Greenwood Press

Creveld, M. van (1977), *Supplying War; Logistics from Wallenstein to Patton*. Cambridge: Cambridge University Press

Creveld, M. van (1986), 'On Learning from the Wehrmacht and Other Things'. *Military Review*, January

Dunigan, J.F. (ed.) (1978), *The Russian Front; Germany's War in the East 1941–1945*. London: Arms and Armour Press

Dupuy, T.N. (1977), *A Genius for War; The German Army and the General Staff*. London: Macdonald & Jane's

Franz, P. (1984), 'Operational concepts'. *Military Review*, July, 2–15

Garder, M. (1966), *A History of the Soviet Army*. London: Pall Mall Press

Glantz, D.M. (2005), *Colossus Reborn; The Red Army at War, 1941–1943*, Lawrence: University Press of Kansas

Griffith. R. (1992), *Forward into Battle*. Navato (Ca): Presidio Press

Guderian, H. (1952/2000), *Panzer Leader*. London: Penguin Group

Habeck, M.H. (2003), *Storm of Steel; the development of Armor Doctrine in Germany and the Soviet Union, 1919–1939*. New York: Cornell University Press

Holden Reid, B. (2000), *A Doctrinal Perspective, 1988–1998*. London: Strategic and Combat Studies Institute

Laffin, J. (1965), *Jackboot; the Story of the German Soldier*. New York: Cassell

Lewin, R. (ed.) (1984), *The War on Land 1939–1945*. London: Random House

Lewis, S.J. (1988), 'Reflections on German Military Reform', *Military Review*, August

Mackenzie, J.J.G. and Holden Reid, B. (eds) (1989), *The British Army and the Operational Level of War*. London: Tri-Service Press Ltd

Messenger, C. (1976), *The Art of Blitzkrieg*. Shepperton (Surrey): Ian Allen Ltd

Millett, A.R. and Murray, W. (eds) (1988), *Military Effectiveness. Volume III, The Second World War*. Boston: Mershon Centre, The Ohio State University

Murray, W. (1990), 'Force strategy, Blitzkrieg strategy and the economic difficulties: Nazi grand strategy in the 1930s'. *Military Review*, May

Murray, W. (1992), *German Military Effectiveness*. New York: The Nautical & Aviation Publishing Company of America

Murray, W and Millet, A.R. (eds) (1996), *Military Innovation in the Interwar Period*. Cambridge: Cambridge University Press

Oeting, D.W. (1993), *Auftragstaktik*. Frankfurt am Main/Bonn: Report Verlag GmbH

Ogilvie, R. (1995), *Krijgen is een kunst*. Amsterdam: Addison Wesley Publishing Company Inc.

Overy, R. (1998), *Why the Allies Won*. New York: W.W. Norton & Company

Perrett, B. (1986/1997), *Knights of the Black Cross*. London: Wordsworth Editions Limited

Posen, B.R. (1984), *The Sources of Military Doctrine; France, Britain and Germany between the two world wars*. New York: Cornell studies in security affairs

Richey, S.W. (1984), 'The Philosophical Basis of the Air Land Battle'. *Military Review*, May, 51–2

Rosinski, H. and Craig, G.A. (eds) (1966), *The German Army*. London: Pall Mall Press

Samuels, M. (1992), *Doctrine and Dogma; German and British Infantry Tactics in the First World War*. New York: Greenwood press

Strachan, H. (1997), *The Politics of the British Army*. Oxford: Clarendon Press

Wilbeck, C.W. (2004), *Sledgehammer; Strengths and Flaws of Tiger Tank Battalions in World War II*. London: Tauris & Co

Wilt, A.F. (1990), *War from the Top; German and British military decision making during World War II*. London: Tauris & Co

Ziemke, E.F. (2004), *The Red Army 1918–1941: from Vanguard of World Revolution to US Ally*. London and New York: Frank Cass

Index